Lars Kadison   Matthias T. Kromann

Projective Geometry
and Modern Algebra

Birkhäuser
Boston · Basel · Berlin

Lars Kadison
Department of Mathematics
University of Copenhagen
2100 Copenhagen
Denmark

Matthias T. Kromann
Department of Mathematics
University of Pennsylvania
Philadelphia, PA 19104
U.S.A.

Library of Congress Cataloging-in-Publication Data

Kadison, Lars,
  Projective geometry and modern algebra / Lars Kadison and Matthias
T. Kromann.
      p.   cm.
  Includes bibliographical references (p. -) and index.
  ISBN 0-8176-3900-4 (hardcover : acid free). -- ISBN 3-7643-3900-4
(hardcover : acid free)
    1. Geometry, Projective.   I. Kromann, Matthias, T.   II. Title.
  QA471.K224   1996
  516'.5--dc20   95-47840
                                                               CIP

Printed on acid-free paper
© 1996 Lars Kadison
Based on *Foundations of Projective Geometry* ,
(W. A. Benjamin, Inc., 1967) © by Robin Hartshorne   *Birkhäuser*

ISBN 0-8176-3900-4
ISBN 3-7643-3900-4
Typeset by the authors
Figures were programmed in PostScript using a PostScript library
developed by Matthias T. Kromann
Printed and bound by Maple-Vail, York, PA
Printed in the U.S.A.

9 8 7 6 5 4 3 2 1

Contents

a lattice of subspaces. They are partly intended to develop several themes in the book. For the most part, though, they are intended to guide the reader through an independent study or an accredited project. Each appendix proposes a central problem that the reader may take up on his own, or pursue in the literature of projective geometry, algebraic geometry, metric geometry or foundations of geometry.

Certain exercises are important and should be done with care. They are marked with asterisks, and their solutions are sketched in the back of the book.

Acknowledgements.

We thank R. Hartshorne, J. Gonzalo and B. Booß-Bavnbek for very helpful discussions about our book. Also our appreciation to the students of E5 for working their way through an early draft of this book.

This book is based on Robin Hartshorne's Foundations of Projective Geometry, W. A. Benjamin, Inc., 1967, a book we admired and were sorry to see out of print. We follow Hartshorne's book closely in Chapters 1-5, and more loosely in Chapters 6-11. The mathematical content is basically the same, except that in Chapter 6 we have taken the Fundamental Theorem as an axiom, and from it have proved Pappus's theorem, while in Hartshorne's book it is the reverse. The main changes from [Hartshorne] are in Sections 3.1, 3.2, 8.3, 11.5, and Chapters 6 and 7. We have also added the Historical Foreword, most of the exercises, and Appendices A-E. We thank R. Hartshorne for written permission to use material from his book.

The authors

Historical Foreword

The motivation for the modern theory of projective geometry came from the fine arts. By 1300 artists like Duccio and Giotto were no longer content with the highly stylized medieval art and sought to revive Graeco-Roman standards: they made the first experiments in foreshortening and the use of converging lines to give an impression of depth in a scene painting. Their intuitive theory of perspective culminated in the work of Lorenzetti in the 1340s.

At this stage, further progress in the realistic representation of 3-dimensional scenes on a 2-dimensional canvas had to await the development of a mathematical theory of perspective. The Italian painter and architect Brunelleschi was teaching such a theory in 1425. In 1435, L. B. Alberti had written the first treatise on the mathematical model for perspective drawing. Later, the gifted mathematician and painter Piero della Francesca (c. 1418–1492) considerably extended the work of Alberti. Still later, both Leonardo da Vinci (1452–1519) and Albrecht Düürer (1471–1528) wrote treatises which not only presented the mathematical theory of perspective but insisted on its fundamental importance in all of painting. In the 1700's, J. H. Lambert and B. Taylor wrote treatises on perspectivity which treated complicated problems such as scenes that include mirrors.

In a separate development pre-dating the theory of perspective by a couple of millenia, Appollonius, Archimedes, Euclid and Menaechmus, in the period from 400 until 200 B.C., had introduced and studied the subject of conics. However, the first truly projective theorems were discovered by Pappus of Alexandria around 250 A.D. (see Chapter 6) and proven by him using complicated Euclidean argumentation.

The German astronomer Johann Kepler (1571–1630) elevated the ellipse of Apollonius to center stage in scientific history with his first law of planetary motion. In 1639 the sixteen year old Blaise Pascal wrote *Essai pour les Coniques*[1], in which he deduced 400 propositions on conics, including the work of Appollonius and others, from the theorem that now bears his name.

---

[1] Now lost but apparently Leibniz had read it.

Figure 1. Print from Jan Vredeman de Vries's book on perspective, 1604.

The proof of Pascal's theorem used the method of projection, which he had learnt from Girard Desargues (1591–1661). The architect and gifted mathematician Desargues added a great body of work to projective geometry, including his two truly great theorems (in Chapter 3 and Appendix A). His work was not well received in his lifetime, which perhaps was due to his obscure style: of seventy terms he introduced, but one (involution) survives today. However, E.T. Bell, in his vigorous biographical style, notes the following irony of history and the passage of time: Bell traces the mathematics of Einstein's general theory of relativity back to Desargues [Bell, p. 213], who was unknown to Isaac Newton (1642–1728).

Newtonian mechanics and calculus had dominated mathematics, physics and philosophy for a century, when a young engineering officer J.V. Poncelet (1788-1867) was facing internment in a prisoner-of-war camp for prisoners taken from Napoleon's Grand Army. He had a solid education in geometry from Monge and the elder Carnot, and set about trying to recall what he had learnt from them. Finding that he could recreate the general principles but could not recall the barren details of the eighteenth century masters, he proceeded to invent projective geometry as we know it today. Among many things, he is the first to have applied the principle of duality in a treatise on projective geometry: his habit of writing about projective geometry in two columned pages, one column for the new theory, the other for its dual, continued into the 20th century.

Projective geometry came into its own as a research field of mathematics after the publication of Poncelet's work. K.G.C. von Staudt (1798–1867) studied conics, polarities, the fundamental theorem, and emancipated the ideal points from their special status. The Swiss Jakob Steiner (1796–1863) studied conics from the point of view of 1-dimensional projectivities. The French M. Chasles (1793–1880) emphasized the cross ratio in his study of conics. Arthur Cayley derived in 1859 (perhaps with a hint from Laguerre's earlier book) the three geometries of Euclid, Bolyai, J. and Lobatchevsky from cross ratio, a fixed conic and region in the real projective plane. Matrix multiplication itself seems to have grown out of Cayley's investigations of the projective invariants of A.F. Möbius (1790–1868). To Möbius and Feuerbach we owe homogeneous coordinates in 1827, but it was left to Felix Klein in 1871 to remove the last vestiges of Euclidean geometry and provide the algebraic foundations of projective geometry that is evident in Chapter 8 of the present book.

In 1899 David Hilbert (1862–1943) published his *Grundlagen der Geometrie* in which the fruit of about 20 years intellectual labor of himself, Pasch and others were recorded. This book can be viewed as the work that set Euclid straight after 2000 years unquestioned intellectual hegemony. Six primitive notions and 20 axioms arranged in five groupings were given for 3-dimensional Euclidean geometry in an almost flawless postulational system. Hilbert showed his postulational system to be as consistent as the arithmetic of real numbers. (In chapters 1–9 of this book, we show by methods similar to Hilbert's that our seven axioms of planar projective geometry are as consistent as the theory of fields of characteristic $\neq 2$.) Later, Hilbert was emboldened to formulate his doctrine of formalism. It is at this time that insights into the foundations of projective geometry were made as well.

As we come into the 20th century we approach a tangle of events that are hard to sort through. Further contributions to projective geometry have been made by E. Artin, R. Baer, G. Birkhoff, J. von Neumann, and O. Veblen. Particularly much work was done in clarifying and generalizing the fundamental theorem (Appendix E). A particular passage in Bell, writing in the '30s, comes to mind:

> *"The conspicuous beauty of projective geometry and the supple elegance of its demonstrations made it a favorite study with the geometers of the 19th century. Able men swarmed into the new goldfield and quickly stripped it of its more accessible treasures. Today the majority of experts seem to agree that the subject is worked out so far as it is of interest to professional mathematicians. However, it is conceivable that there may be something in it as obvious as the principle of duality which has been overlooked."* [2]

Indeed Marshall Hall's 1943 article where he coordinatizes projective planes with ternary rings — and the subsequent charting of non-Desarguesian geometry,

---

[2]Bell continues encouragingly: "In any event it is an easy subject to acquire and one of fascinating delight to amateurs and even to professionals at some stage of their careers."

finite or not — would probably qualify in Bell's own opinion as a brilliantly obvious development (see Appendix D). New projective geometry continues up to the present day (e.g., [Holland]).

Before we leave the history of our subject, and take up its close study, we would like to identify our text and its place in this large mosaic. According to Philip Davis, a course very similar to ours was taught for many years at Harvard University by Oscar Zariski, a well-known algebraic geometer who emphasized the role of commutative rings in this subject. Robin Hartshorne gave a similar course at Harvard in the '60s and wrote up his lecture notes in [Hartshorne]. The senior author is indebted to [Hartshorne], from which he learned the present subject himself, before teaching it several times in the same tradition: hearty thanks to Gestur Ólafsson for recommending these notes to him.

Chapter 1

Affine Geometry

Projective geometry developed from considerations of geometric properties invariant under central projection. Properties of incidence such as "collinearity of points", "concurrence of lines" and "triangularity" are invariants under central projection, while ordinary notions of distance, angle, and parallelism are noninvariants, as they are visibly distorted under central projection. Thus in the axiomatic development of the theory our focus is on properties of incidence without parallelism.

However, one of the most important examples of the theory is the real projective plane. There we will use all techniques available to us from Euclidean geometry and analytic geometry in order to see what is true or not true.

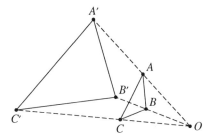

Figure 1.1. Central projection with center O.

As the center of projection of one plane to another moves far enough away we become concerned with invariants of figures under parallel projection. In this limiting case where O in Figure 1.1 is taken to be infinitely far away, parallel lines project to parallel lines as one may see by taking different planar sections of two parallel planes. Let us then start out on our path toward projective geometry with some of the most elementary facts about incidence in the ordinary plane geometry, which we take as axioms for our synthetic development. In Chapter 2, we will

1

point out the very close relationship of affine plane to projective plane, and we will exploit this relation throughout the text.

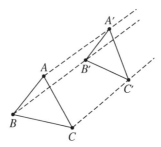

Figure 1.2. Parallel projection.

1.1   Affine Planes

The following definition is a distillation of the most basic properties of incidence in the Euclidean plane. We will then see that our definition includes some surprising examples.

1.1. Definition   An *affine plane* is a set, whose elements are called *points*, and a set of subsets, called *lines*, satisfying the following three axioms, A1–A3. We will use the terminology "*P* lies on ℓ" or "ℓ passes through *P*" to mean the point *P* is an element of the line ℓ.

> **A1.** Given two distinct points *P* and *Q*, there is one and only one line containing both *P* and *Q*.
> **A2.** Given a line ℓ and a point *P*, not on ℓ, there is one and only one line *m*, having no point in common with ℓ, and passing through *P*.
> **A3.** There exist three noncollinear points.

A set of points $P_1, \ldots, P_n$ is said to be *collinear* if there exists a line ℓ containing them all. We say that two lines are *parallel* if they are equal, or if they have no points in common.

1.2. Notation   Throughout the book, we use the following notation.

$X \smallsetminus B$	the complement of *B* in *X*
$\#(S)$	number of elements in a finite set *S*
$P \neq Q$	*P* is not equal to *Q*
$P \in \ell$	*P* lies on ℓ
$PQ, P \cup Q$	line through *P* and *Q*

$\ell.m$, $\ell \cap m$	the intersection of ℓ and m
$\ell.m = P$	ℓ and m intersect in P
$\ell \parallel m$	ℓ is parallel to m
$\forall$	for all
$\exists$	there exists
$\Longrightarrow$	implies
$\Longleftrightarrow$	if and only if (iff)
$\emptyset$	the empty set

1.3. Example The ordinary plane, known to us from Euclidean geometry, satisfies the axioms A1–A3, and therefore is an affine plane, the *real affine plane*.

A convenient way of representing this plane is by introducing Cartesian coordinates, as in analytic geometry. Thus a point P is represented as an ordered pair (x,y) of real numbers. A line is the set of solutions (x,y) of a linear equation $y = mx + b$ or $x = a$.

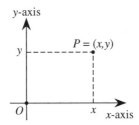

Figure 1.3. The real affine plane.

1.4. Definition A relation $\mathfrak{R}$ on a set S is a subset of the cartesian product $S \times S$ of ordered pairs of elements from S. If $(a,b) \in \mathfrak{R}$ one usually writes $a\mathfrak{R}b$. A relation $\sim$ on a set $S = \{a,b,c,\dots\}$ is an *equivalence relation* if it has the following three properties:

    (1) It is *reflexive*: $a \sim a$.
    (2) It is *symmetric*: If $a \sim b$, then $b \sim a$.
    (3) It is *transitive*: If $a \sim b$ and $b \sim c$, then $a \sim c$.

The *equivalence class* $[a]$ of a is the subset of elements equivalent to a:

$$[a] = \{b \in S \mid b \sim a\}.$$

1.5. Proposition *Parallelism is an equivalence relation.*

Proof We must check the three properties.

   (*1*) Any line is parallel to itself, by definition.

   (*2*) If $\ell \parallel m$, then $m \parallel \ell$ by definition.

   (*3*) If $\ell \parallel m$ and $m \parallel n$, we wish to prove $\ell \parallel n$. If $\ell = n$, there is nothing to prove. If $\ell \neq n$, and there is a point $P \in \ell \cap n$, then ℓ, n are both parallel to m, and pass through P, which is impossible by Axiom A2. We conclude that $\ell \cap n = \varnothing$, and so $\ell \parallel n$.   □

1.6. Proposition *Two distinct lines have at most one point in common.*

Proof For if ℓ and m both pass through two distinct points P and Q, then $\ell = m$ by Axiom A1.   □

1.7. Proposition *An affine plane has at least four points.*

Proof Indeed, by A3 there are three noncollinear points. Call them P, Q, R. By A2 there is a line ℓ through P, parallel to the line QR, which exists by A1. Similarly, there is a line $m \parallel PQ$, passing through R.

   Now, ℓ is not parallel to m (we write $\ell \nparallel m$). For if it were, then we would have $PQ \parallel m \parallel \ell \parallel QR$ and hence $PQ \parallel QR$ by Proposition 1.5. This is impossible, however, because $PQ \neq QR$, and both contain Q.

   Hence ℓ must meet m in some point S. Since S lies on m, which is parallel to PQ, and different from PQ, S does not lie on PQ, so $S \neq P$, and $S \neq Q$. Similarly $S \neq R$. Thus S is indeed a fourth point.   □

1.8. Example There is an affine plane with four points.

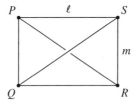

Figure 1.4. A pictorial representation of the four point affine plane.

   We have an affine plane consisting of four points $\{P, Q, R, S\}$, and six lines $\{P, Q\}$, $\{P, R\}$, $\{P, S\}$, $\{Q, R\}$, $\{Q, S\}$, $\{R, S\}$: one easily verifies the axioms A1–A3 (Exercise 1.1). This is the smallest affine plane by Proposition 1.7.

1.9. Definition A *pencil of lines* is either

   (*1*) the set of all lines passing through some point P, or

(2) the set of all lines parallel to some line ℓ.

In the second case we speak of a *pencil of parallel lines*.

1.10. Definition A *one-to-one correspondence* between two sets X and Y is a mapping $T\colon X \to Y$ (i.e. a rule T, which associates to each element x of the set X one element $T(x) = y$ in Y) such that

(1) $x_1 \neq x_2$ implies that $Tx_1 \neq Tx_2$, and such that
(2) for all $y \in Y$ there exists $x \in X$ such that $T(x) = y$.

A one-to-one correspondence of a set X with itself is called a *permutation* of X.

1.2   Transformations of the Affine Plane

Any other labelling of the 4 points in Figure 1.4 with the letters P, Q, R, and S induces a permutation of $\{P,Q,R,S\}$ that sends lines to lines and preserves parallelism. For example, exchanging P and Q sends lines PR and QS to lines QR and PS, respectively. However, the resulting sets of points and lines is the same as before. In addition, a new labelling of points in the real affine plane coming from a change in coordinate axes is itself a permutation of $\mathbf{R}^2$, given by

$$\begin{pmatrix} x' \\ y' \end{pmatrix} = \begin{pmatrix} a_{11} & a_{12} \\ a_{21} & a_{22} \end{pmatrix} \begin{pmatrix} x \\ y \end{pmatrix} + \begin{pmatrix} b_1 \\ b_2 \end{pmatrix}$$

where $a_{11}a_{22} - a_{12}a_{21} \neq 0$; or in more compact matrix and vector notation we could write $v' = Av + b$ where $\det A \neq 0$. A is the transition matrix between old and new bases.

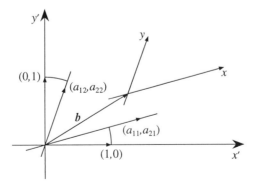

Figure 1.5. Transformation to new coordinates in the real affine plane.

Now, true affine invariants should not depend on the labelling of points in a representation of an affine plane. In the next definition we give the precise meaning of "relabelling a geometry."

1.11. Definition Let **A** be an affine plane. An *automorphism* ϕ of **A** is a permutation of **A** sending collinear points into collinear points.

1.12. Convention We will often say an automorphism ϕ transforms a point P to P', point Q to Q', line ℓ to ℓ', i.e. $\phi(P) = P'$, $\phi(Q) = Q'$, $\phi(\ell) = \ell'$, so priming a lettered point or line is used to denote the image point or line under ϕ. For example, if P, Q, R are collinear points, an automorphism ϕ sends these into collinear points P', Q', R', according to the definition.

In addition, different notations stand for distinct elements: $P, Q \in$ **A** will mean $P \neq Q$ unless stated otherwise. Transformation or symmetry are older synonyms for automorphism.

1.13. Remark It follows from elementary considerations that given a line $\ell = PQ$, ϕ restricts to a one-to-one correspondence between the sets of points ℓ and ℓ'.

We observe that an automorphism satisfies the two "algebraic" conditions

(1) $\phi(P \cup Q) = \phi(P) \cup \phi(Q)$ ($\forall P, Q \in$ **A**)

(2) $\phi(\ell.m) = \phi(\ell).\phi(m)$ ($\forall \ell, m : \ell \nparallel m$)

1.14. Proposition *An automorphism transforms parallel lines to parallel lines.*

Proof Suppose $PQ \parallel RS$ in an affine plane **A**. If $B' \in P'Q' \cap R'S'$, then, by the remark, $B \in PQ \cap RS$, so $PQ = RS$. Hence $P'Q' = R'S'$. If $P'Q'$ and $R'S'$ had no point in common, they would be parallel, leaving nothing more to prove. $\quad\square$

1.15. Proposition *The set of automorphisms* Aut**A** *is closed under composition and inversion.*

Proof We would like see that $\phi, \psi \in$ Aut**A** imply $\phi \circ \psi \in$ Aut**A** and $\phi^{-1} \in$ Aut**A**. Now both $\phi \circ \psi$ and ϕ^{-1} are permutations of **A**. We check lines go to lines.

If P, Q, and R are collinear, then $\psi(P)$, $\psi(Q)$, and $\psi(R)$ are collinear, and so are $\phi(\psi(P))$, $\phi(\psi(Q))$, and $\phi(\psi(R))$. Also, $P = \phi(A)$, $Q = \phi(B)$, and $R = \phi(C)$ for three points $A, B, C \in$ **A**. A, B, and C are collinear for otherwise $AB \neq BC$, so $PQ \neq QR$ as lines. But $A = \phi^{-1}(P), B = \phi^{-1}(Q)$, and $C = \phi^{-1}(R)$ so ϕ^{-1} also takes lines to lines. $\quad\square$

We now study some special automorphisms in affine geometry that are needed in Chapter 9.

1.16. Definition Let **A** be an affine plane. A *dilatation* is an automorphism $\phi\colon x \mapsto x'$ of **A**, such that for any two distinct points P, Q,

$$PQ \parallel P'Q'.$$

In other words, ϕ takes each line into a parallel line.

1.17. Examples In the *real affine plane* $\mathbf{A}^2(\mathbf{R}) = \{\,(x,y) \mid x,y \in \mathbf{R}\,\}$, a *stretching* in the ratio k, given by equations

$$(x,y) \mapsto (x',y')\colon \begin{cases} x' = kx \\ y' = ky \end{cases}$$

is a dilatation. Indeed, let O be the point $(0,0)$ then ϕ stretches points away from O by a factor of k. If P,Q are any two points, then $PQ \parallel P'Q'$, by similar triangles.

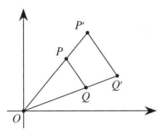

Figure 1.6. Stretching.

Another example of a dilatation of $\mathbf{A}^2(\mathbf{R})$ is given by a translation

$$\begin{cases} x' = x + a \\ y' = y + b \end{cases}$$

In this case, any point P is translated by the vector (a,b), so $PQ \parallel P'Q'$ again, for any P,Q.

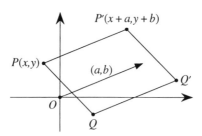

Figure 1.7. Translation.

Without asking for the moment whether there *are* any nontrivial dilatations in a given affine plane **A**, let us study some of their properties.

1.18. Proposition *Let* **A** *be an affine plane. Then the set of dilatations,* Dil **A***, is closed under composition and inversion.*

Proof Indeed, we must see that the product of two dilatations is a dilatation, and that the inverse of a dilatation is a dilatation. This follows immediately from the fact that parallelism is an equivalence relation. □

1.19. Proposition *A dilatation which leaves two distinct points fixed is the identity.*

Proof Let ϕ be a dilatation, let P, Q be fixed, and let R be any point not on PQ. Let $\phi(R) = R'$. Then we have $PR \parallel PR'$ and $QR \parallel QR'$ since $P = P'$ and $Q = Q'$. Hence $R' \in PR$ and $R' \in QR$. But $PR \neq QR$ since $R \notin PQ$. Hence $PR.QR = R'$, and so $R = R'$, i.e. R is also fixed. But R was an arbitrary point not on PQ. Applying the same argument to P and R, we see that every point of PQ is also fixed, so ϕ is the identity. □

1.20. Corollary *A dilatation is determined by the images of two points, i.e. any two dilatations* ϕ, ψ*, which behave the same way on two distinct points* P, Q*, are equal.*

Proof Indeed, $\psi^{-1}\phi$ leaves P, Q fixed, so is the identity. □

So we see that a dilatation different from the identity can have at most one fixed point. We have a special name for those dilatations with no fixed points:

1.21. Definition A *translation* is a dilatation with no fixed points, or the identity.

1.22. Proposition *If* ϕ *is a translation, different from the identity, then for any two points* P, Q*, we have* $PP' \parallel QQ'$*, where* $\phi(P) = P'$*,* $\phi(Q) = Q'$*.*

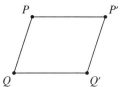

Figure 1.8. Translation: two points and their images form a parallelogram.

Proof Suppose $PP' \nparallel QQ'$. Then these two lines intersect in a point O. But the fact that ϕ is a dilatation implies that ϕ sends the line PP' into itself, and ϕ sends QQ' into itself. (For example, let $R \in PP'$. Then $PR \parallel P'R'$, but $PR = PP'$, so $R' \in PP'$). Hence $\phi(O) = O$, a contradiction.   □

1.23. Proposition *The translations of* **A** *form a subset* Tran **A** *of the set of dilatations of* **A**, *which is closed under composition and inversion. Furthermore, for any* $\tau \in$ Tran **A**, *and* $\sigma \in$ Dil **A**, $\sigma\tau\sigma^{-1} \in$ Tran **A**.

Proof First we must check that the product of two translations is a translation, and the inverse of a translation is a translation. Let τ_1, τ_2 be translations, then $\tau_1\tau_2$ is a dilatation. Suppose it has a fixed point P. If $\tau_2(P) = P'$, then $\tau_1(P') = P$. If Q is any point not on PP', then let $Q' = \tau_2(Q)$.

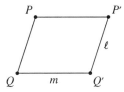

Figure 1.9. Construction of Q' as fourth vertex of parallelogram.

We have $PQ \parallel P'Q'$ and by the previous proposition, $PP' \parallel QQ'$. Hence Q' is determined as the intersection of the line $\ell \parallel PQ$ through P', and the line $m \parallel PP'$ through Q.

For a similar reason, $\tau_1(Q') = Q$. Hence $\tau_1\tau_2$ leave P and Q fixed, so by our proposition $\tau_1\tau_2 = $ id. Hence $\tau_1\tau_2$ is a translation. Clearly the inverse of a translation is a translation, so the translations form a subset of Dil **A**, which is closed under composition and inverse.

Now let $\tau \in$ Tran **A**, $\sigma \in$ Dil **A**. Then $\sigma\tau\sigma^{-1}$ is certainly a dilatation. If it has no fixed points, it is a translation. If it has a fixed point P, then $\sigma\tau\sigma^{-1}(P) = P$ implies $\tau\sigma^{-1}(P) = \sigma^{-1}(P)$, so τ has a fixed point. Hence $\tau = $ id, and $\sigma\tau\sigma^{-1} = $ id. In either case, $\sigma\tau\sigma^{-1}$ is a translation.   □

The question of existence of translations and central dilatations between arbitrary pairs of points is taken up in Chapter 9. The number of automorphisms of the four and nine point affine planes are treated in Exercises 1.6 and 4.10.

Exercises

1.1   Check that the set of points $\{P, Q, R, S\}$ and set of subsets

$$\{\{P,Q\}, \{P,R\}, \{P,S\}, \{Q,R\}, \{Q,S\}, \{R,S\}\}$$

satisfy the axioms A1–A3 of an affine plane.

1.2   Show that any two pencils $[m]$ and $[n]$ of parallel lines in an affine plane have the same cardinality (i.e. that one can establish a one-to-one correspondence between them). Show that this is also the cardinality of the set of points on any line.

1.3   If there is a line with exactly n points, show that the number of points in the whole affine plane is n^2.

1.4   Construct an affine plane with 16 points.

   Hint: We know from Exercise 1.2 that each pencil of parallel lines has four lines in it. Let a, b, c, d be one pencil of parallel lines, and let $1, 2, 3, 4$ be another. Then label the intersections $A_1 = a \cap 1$, etc. To construct the plane, you must choose other subsets of four points to be the lines in the three other pencils of parallel lines. Write out each line explicitly, by naming its four points, e.g. $2 = \{A_2, B_2, C_2, D_2\}$.

1.5   Euler in 1782 posed the following problem: "A meeting of 36 officers of six different ranks and from six different regiments must be arranged in a square in such a manner that each row and each column contains 6 officers from different regiments and of different ranks." G. Tarry in 1901 confirmed Euler's prediction that this problem has no solution. Deduce from Tarry's fact that there is no affine plane with 36 points.

1.6   Given the list of lines $\{P,R\}, \{P,S\}, \{P,Q\}, \{Q,S\}, \{Q,R\}, \{R,S\}$, how many ways consistent with this list are there to label the points in Figure 1.4 with the letters P, Q, R, and S? How many automorphisms of this affine plane are there?

1.7   Which relations below are equivalence relations?

   (a) $\geq$ on the set of integers;
   (b) "is similar to" on the set of triangles of the Euclidean plane;
   (c) "is divisible by" on the set of natural numbers.

1.8   A *partition* of a set S is a set of subsets of S, $\{X_1, \ldots, X_n, \ldots\}$, such that $\bigcup X_i = S$ and $X_i \cap X_j = \emptyset$ whenever $i \neq j$.

   (a) Show that the set of equivalence classes of a set S is a partition of S, given some equivalence relation on S.

   (b) Conversely, show that a partition of S naturally determines an equivalence relation on S.

1.9   Given a 2×2 real matrix A with nonzero determinant and 2×1 real column vector $\boldsymbol{b}$, define an *affine transformation* T to be the bijection of $\mathbf{R}^2$ with itself given by $T(\boldsymbol{x}) = A\boldsymbol{x} + \boldsymbol{b}$. Show that T is an automorphism of the real affine plane.

1.10   A *shear* T is an affine transformation taking Q to R, fixing every point on another line ℓ parallel to QR. By translating and changing basis, find a matrix A and vector $\boldsymbol{b}$ such that $T(\boldsymbol{x}) = A\boldsymbol{x} + \boldsymbol{b}$.

1.11   An *affine reflection* is an affine transformation T that interchanges two points Q and R and fixes a third point P not on QR. Show that T fixes every point on PM, where M is the midpoint of the line segment QR.

1.12   An *isometry* T of the Euclidean plane is a mapping of $\mathbf{R}^2$ onto itself that preserves distance: i.e. if $\mathrm{dist}(P,Q)$ denotes the Euclidean distance from P to Q, then

$$\mathrm{dist}(T(P), T(Q)) = \mathrm{dist}(P, Q) \qquad (\forall P, Q).$$

Show that T sends lines to lines.

1.13   Suppose τ is a translation of an affine plane $\mathbf{A}$ such that $\tau(P) = P'$. Suppose Q is a point not on the line PP'. Show that $Q' = \tau(Q)$ may be obtained by the following parallelogram construction:
Let $\ell \parallel PP'$ such that $Q \in \ell$. Let $m \parallel PQ$ such that $P' \in m$. Then $Q' = \ell.m$.

1.14   Suppose A, B, C are three collinear points in the real affine plane. Define the quantity

$$(A, B; C) = \frac{AC}{BC}$$

where AC the signed Euclidean length of line segment $\overline{AC}$ (e.g., $AC = -CA$).
(*a*) Show that C is the midpoint of the line segment $\overline{AB}$ iff $(A, B; C) = -1$.
(*b*) If A', B', C' are three collinear points, which are obtained as images of collinear points A, B, C, under parallel projection, show that $(A', B'; C') = (A, B; C)$.

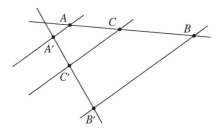

Figure 1.10. The affine invariant $(A, B; C)$ of three collinear points.

Chapter 2

Projective Planes

There are two points of view as to why we should avoid parallel lines in planar geometry. One is that lines and indeed higher degree polynomial curves should intersect in the "right" number of points: two lines, for instance, usually intersect in one point and should therefore do so at all times. This leads to the introduction of a common "ideal point" to each line in a pencil of parallels.

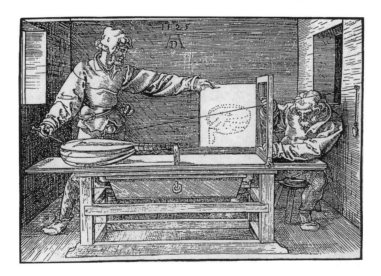

Figure 2.1. Print from the 1525 edition of Albrecht Dürer's book on perspective.

The second point of view stems from the art of perspective which holds that a 2- or 3-dimensional figure should be united with the eye of the observer O by

13

lines to produce a "cone" over the figure. A drawing of the figure corresponds to a planar cross section of this cone. For example, the cone over parallel lines is two planes intersecting in a line through O, cross sections of which generally are two intersecting lines. This model for the art of perspective leads naturally to homogeneous coordinates for the real projective plane.

We will show in Proposition 2.9 that the two points of view on the real projective plane are equivalent in a certain strictly defined sense.

2.1   Completion of the Affine Plane

We complete the affine plane by adding certain "points at infinity" and thus arrive at the notion of the projective plane.

Let $\mathbf{A}$ be the affine plane. For each line $\ell \in \mathbf{A}$, we denote by $[\ell]$ the pencil of lines parallel to ℓ. To each pencil of parallels $[\ell]$ we add to $\mathbf{A}$ an *ideal point*, or *point at infinity in the direction of* ℓ, which we denote by $P_{[\ell]}$.

We define the *completion* $\mathbf{S}$ of $\mathbf{A}$ as follows. The *points* of $\mathbf{S}$ are the points of $\mathbf{A}$, with the addition of all the ideal points of $\mathbf{A}$. A *line* in $\mathbf{S}$ is either

    (*a*) an ordinary line ℓ of $\mathbf{A}$, plus the ideal point $P_{[\ell]}$ of ℓ, or

    (*b*) the *line at infinity*, consisting of all the ideal points of $\mathbf{A}$.

We will see shortly that $\mathbf{S}$ is a projective plane, in the sense of the following definition.

2.1. Definition A *projective plane* $\mathbf{P}$ is a set, whose elements are called points, and a set of subsets, called lines, satisfying the following four axioms.

    P1. Two distinct points P, Q of $\mathbf{P}$ lie on one and only one line.

    P2. Two distinct lines meet in precisely one point.

    P3. There exist three noncollinear points.

    P4. Every line contains *at least* three points.

2.2. Proposition *The completion* $\mathbf{S}$ *of an affine plane* $\mathbf{A}$, *as described above, is a projective plane.*

    Proof We must verify the four axioms P1–P4 of the definition.

    P1. Let $P, Q \in \mathbf{S}$. (*1*) If P, Q are ordinary points of $\mathbf{A}$, then P and Q lie on only one line of $\mathbf{A}$. They do not lie on the line at infinity of $\mathbf{S}$, hence they lie on only one line of $\mathbf{S}$. (2) If Q is an ordinary point, and $P_{[\ell]}$ is an ideal point, we can find by A2 a line m such that $Q \in m$, and $m \parallel \ell$, i.e. $m \in [\ell]$, so that $P_{[\ell]}$ lies on the extension of m to $\mathbf{S}$. This is clearly the only line of $\mathbf{S}$ containing

$P_{[\ell]}$ and Q. (*3*) If P,Q are both ideal points, then they both lie on the line "at infinity", the only line of **S** containing them.

P2. Let ℓ, m be lines. (*1*) If they are both ordinary lines, and $\ell \nparallel m$, then they meet in a point of **A**. If $\ell \parallel m$, then the ideal point $P_{[\ell]}$ lies on both ℓ and m in **S**. (*2*) If ℓ is an ordinary line, and $m = \ell_\infty$ is the line at infinity, then $P_{[\ell]}$ lies on both ℓ and m.

P3. Follows immediately from A3. One must check only that if P,Q,R are noncollinear in **A**, then they are also noncollinear in **S**. Indeed, the only new line is the line at infinity, which contains none of them.

P4. It is easy to see that each line of **A** contains at least two points. But in **S** it has also an ideal point, so any line has at least three points. (It follows from Exercise 1.2 that ℓ_∞ has at least three points).   □

2.3. Example By completing the real affine plane of Euclidean geometry, we obtain the real projective plane. Since parallel lines in the real affine plane have equal slope – and conversely – we might identify the "new" ideal points by a slope coordinate m, where $-\infty < m \le \infty$. For example, the vertical lines, i.e., lines parallel to the y-axis, meet at the point $m = \infty$ in the real projective plane.

2.4. Example By completing the affine plane of 4 points, we obtain a projective plane with 7 points. See the figure below.

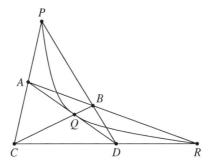

Figure 2.2. A pictorial representation of the seven point projective plane.

2.5. Example Another example of a projective plane can be constructed as follows: let $\mathbf{R}^3$ be ordinary Euclidean 3-space, and let O be a point of $\mathbf{R}^3$. Let **S** be the set of lines through O. We define a *point* of **S** to be a line through O in $\mathbf{R}^3$. We define a *line* of **S** to be the collection of lines through O which all lie in some plane through O. Then **S** satisfies the axioms P1–P4 (Exercise 2.7), and so it is a projective plane.

2.6. Example Let V be a three dimensional vector space over some scalar field.

Consider the set of one dimensional subspaces of V and the set of subsets of these given by the two dimensional subspaces of V. These are the points and lines, respectively, of a projective plane. If the field of scalars is the real numbers, we are just re-stating Example 2.5 in terms of elementary linear algebra.

2.7. The Relationship of Affine Planes to Projective Planes. Completion of an affine plane may be viewed as a function that associates to each affine plane a projective plane: denote it by $\mathbf{A} \to S(\mathbf{A})$. In Exercise 2.2 you will show that to each projective plane $\mathbf{P}$ and line ℓ there is an affine plane $\mathbf{P} \smallsetminus \ell$, which is the complement of ℓ in $\mathbf{P}$ as a set of points, where a line is a line of $\mathbf{P}$ with the point in ℓ deleted. In the exercise, you will also show that $S(\mathbf{P} \smallsetminus \ell)$ is isomorphic to $\mathbf{P}$ (no matter which line ℓ you start with). Conversely, completing an affine plane, then deleting the line at infinity leaves us with the affine plane we started with: $S(\mathbf{A}) \smallsetminus \ell_\infty = \mathbf{A}$. In Chapter 9, we will see that for "reasonable" affine planes, $S(\mathbf{A}) \smallsetminus \ell \cong \mathbf{A}$ for any line ℓ. On this class of "reasonable" affine planes and the associated class of projective planes, the functions $\mathbf{A} \mapsto S(\mathbf{A})$ and $\mathbf{P} \to \mathbf{P} \smallsetminus \ell$ are inverses to one another, up to isomorphism. It is this close relationship between affine planes and projective planes that we will come to understand better through the application of Proposition 2.2 and Exercise 2.2 to many problems in projective geometry.

2.2 Homogeneous Coordinates for the Real Projective Plane

We can give an analytic definition of the real projective plane as follows. We consider Example 2.5 given above of lines in $\mathbf{R}^3$. A point of $\mathbf{S}$ is a line p through the origin O. We will represent the point P of $\mathbf{S}$ corresponding to p by choosing any point (x_1, x_2, x_3) on p different from the point $(0,0,0)$. The numbers x_1, x_2, x_3 are *homogeneous coordinates* of P. Any other point of p has the coordinates

$$(\lambda x_1, \lambda x_2, \lambda x_3),$$

where $\lambda \in \mathbf{R}$, $\lambda \neq 0$. Thus $\mathbf{S}$ is the collection of triples (x_1, x_2, x_3) of real numbers, not all zero, and two triples (x_1, x_2, x_3) and (x'_1, x'_2, x'_3) represent the same point if and only if there exists $\lambda \in \mathbf{R} \smallsetminus \{0\}$ such that $x'_i = \lambda x_i$ for $i = 1, 2, 3$. Since the equation of a plane in $\mathbf{R}^3$ passing through O is of the form

$$a_1 x_1 + a_2 x_2 + a_3 x_3 = 0 \qquad (\text{not all } a_i = 0)$$

we see that this is also the equation of a line of $\mathbf{S}$, in terms of the homogeneous coordinates.

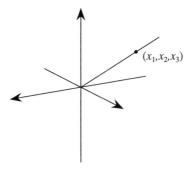

Figure 2.3. Homogeneous coordinates.

2.8. Definition Two projective planes **S** and **S**$'$ are *isomorphic* if there exists a one-to-one correspondence $T: \mathbf{S} \to \mathbf{S}'$ which takes collinear points into collinear points. T is referred to as an *isomorphism*; or as an *automorphism* if $\mathbf{S}' = \mathbf{S}$.

2.9. Proposition *The projective plane* **S** *defined by homogeneous coordinates of real numbers, as above, is isomorphic to the projective plane obtained by completing the ordinary affine plane of Euclidean geometry.*

> **Proof** On the one hand, we have **S**, whose points are given by homogeneous coordinates (x_1, x_2, x_3), $x_i \in \mathbf{R}$, not all zero. On the other hand, we have the Euclidean plane **A**, with Cartesian coordinates (x, y). Let us call the completion **S**$'$. Thus the points of **S**$'$ are the points (x, y) of **A** (with $x, y \in \mathbf{R}$), plus the ideal points. Now a pencil of parallel lines is uniquely determined by its slope m, which may be any real number, or ∞. Thus the ideal points are described by coordinate m.
>
> Now we will define a mapping $T: \mathbf{S} \to \mathbf{S}'$ which will exhibit the isomorphism of **S** and **S**$'$. Let $(x_1, x_2, x_3) = P$ be a point of **S**.
>
> (*1*) If $x_3 \neq 0$, we define $T(P)$ to be the point of **A** with coordinates
>
> $$x = x_1/x_3, \qquad y = x_2/x_3.$$
>
> Note that this is uniquely determined, because if we replace (x_1, x_2, x_3) by $(\lambda x_1, \lambda x_2, \lambda x_3)$, then x and y do not change. Note also that every point of **A** can be obtained in this way. Indeed, the point with coordinates (x, y) is the image of the point of **S** with homogeneous coordinates $(x, y, 1)$.
>
> (*2*) If $x_3 = 0$, then we define $T(P)$ to be the ideal point of **S**$'$ with slope-coordinate $m = x_2/x_1$. Note that this makes sense if we set $x_2/0 = \infty$, because x_1 and x_2 cannot both be zero. Again replacing $(x_1, x_2, 0)$ by $(\lambda x_1, \lambda x_2, 0)$ does not change m. Also each value of m occurs: If $m \neq \infty$, we take $T(1, m, 0)$, and if $m = \infty$, we take $T(0, 1, 0)$.

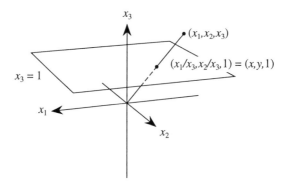

Figure 2.4. The mapping T on points $x_3 \neq 0$.

Thus T is a one-to-one mapping of **S** into **S**′. We must check that T takes collinear points into collinear points. A line ℓ in **S** is given by an equation

$$a_1 x_1 + a_2 x_2 + a_3 x_3 = 0$$

(*1*) Suppose that a_1 and a_2 are not both zero. Then for those points with $x_3 = 0$, namely the point given by $x_1 = \lambda a_2$, $x_2 = -\lambda a_1$, T of this point is the ideal point given by the slope $m = -a_1/a_2$, which indeed is on a line in **S**′ with the finite points.

(*2*) If $a_1 = a_2 = 0$, ℓ has the equation $x_3 = 0$. Any point of **S** with $x_3 = 0$ goes to an ideal point of **S**′, and these form a line. □

2.10. Remark From now on, we will not distinguish between the two isomorphic planes of Proposition 2.9: we will call it the *real projective plane* and denote it by $\mathbf{P}^2(\mathbf{R})$. Notice that the distinction between ordinary points and ideal points in the first model loses its meaning in the second model. For we could just as well have chosen to divide by the x_1 or x_2 coordinate in defining T. Also the distinction between ordinary line and line at infinity loses its meaning.

The real projective plane will be the most important example of the axiomatic theory we are going to develop, and we will often check results of the axiomatic theory in this plane by way of example. Similarly, theorems in the real projective plane can give motivation for results in the axiomatic theory. However, to establish a theorem in our theory, *we must derive it from the axioms and from previous theorems*. If we find that it is true in the real projective plane, that is evidence in favor of the theorem, but does not constitute a proof in our set-up.

Also note that if we remove any line from the real projective plane, we obtain the Euclidean plane. Here is a nice application of that idea. Consider the *conic* in the real projective plane given by $\{ (x,y,z) \mid x^2 + y^2 - z^2 = 0 \}$. Removing the lines $z = 0$, $y = 0$, or $z - y = 0$, we obtain a circle, hyperbola, or parabola (respectively) in the Euclidean plane: these are the *conic sections* of the planes $z = 1$, $y = 1$, and $z - y = 1$, respectively.

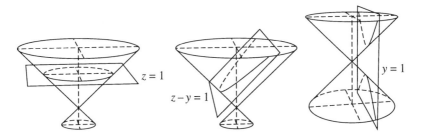

Figure 2.5. Three conic sections: circle, parabola, and hyperbola.

2.11. The sphere model Consider some basic facts about the standard sphere

$$S^2 = \{ (x_1, x_2, x_3) \mid x_1^2 + x_2^2 + x_3^2 = 1 \}.$$

Lines in $\mathbf{R}^3$ through the origin intersect S^2 in antipodal points, and planes through the origin intersect S^2 in great circles. Conversely, antipodal points $\pm(x_1, x_2, x_3)$ on S^2 lie on one line through the origin; and a great circle on S^2 is coplanar with the origin.

Now define a projective plane $\mathbf{S}''$ as follows: a point is a pair of antipodal points on S^2, and a line is a set of points lying on a great circle on S^2. We have just set up an isomorphism between the projective plane of homogeneous coordinates $\mathbf{S}$ and $\mathbf{S}''$. (What is it?) $\mathbf{S}''$ is a third model for the real projective plane.

2.12. Remark The sphere model $\mathbf{S}''$ of the real projective plane is a topological oddity. By considering a hemisphere and pinching antipodal points together on the boundary, one can see how exotic the real projective plane is topologically. It is a nonorientable surface, embeddable in $\mathbf{R}^4$ but not in $\mathbf{R}^3$. It is topologically equivalent to a Möbius strip with a disk attached to its boundary. To see this, remove a cap around the pole of the upper hemisphere, remove the image of cap under the antipode map $v \mapsto -v$ from the lower hemisphere, and visualize the rest of $\mathbf{S}''$ as a circular band with antipodal points identified — now half of this circular band is a band whose ends are placed together with the well-known Möbius twist.

Exercises

2.1 Discuss the possible systems of points and lines which satisfy P1, P2, P3, but not P4.

2.2* Let **S** be a projective plane, and let ℓ be a line of **S**. Define S_0 to be the points of **S** not on ℓ, and define lines in S_0 to be the restrictions of lines in **S**.

(a) Using P1–P4, prove that S_0 is an affine plane.

(b) Prove that **S** is isomorphic to the completion of the affine plane S_0.

2.3 (a) Check that the points $\{A,B,C,D,P,Q,R\}$ with lines

$$\{\{A,C,P\},\{C,D,R\},\{B,D,P\},\{A,B,R\},\{B,C,Q\},\{A,D,Q\},\{P,Q,R\}\}$$

satisfy the axioms P1–P4 for a projective plane.

(b) Prove that the projective plane of 7 points, obtained by completing the affine plane of four points, is the smallest possible projective plane.

(c) Show that two projective planes of 7 points are isomorphic.

2.4* If one line in a projective plane has $n+1$ points, find the number of points in the projective plane.

2.5 (a) Give a proof that the axioms P1, P2, P3, and P4 of a projective plane imply the statement

Q: "There are four points, no three of which are collinear"

(b) Prove also that P1, P2, and Q imply P3 and P4.

2.6 (A) In the real projective plane, what is the equation of the line

(a) joining the points $(1,0,1)$ and $(1,2,3)$?

(b) joining the points $(0,3,9)$ and $(0,25,43)$?

(c) joining the points (a_1,a_2,a_3) and (b_1,b_2,b_3)?

(B) What is the point of intersection

(a) of the lines

$$x_1 - x_2 + 2x_3 = 0$$
$$3x_1 + x_2 + x_3 = 0 \ ?$$

(b) of the lines

$$x_1 + x_3 = 0$$
$$x_1 + 2x_2 + 3x_3 = 0 \ ?$$

(c) of the lines

$$A_1x_1 + A_2x_2 + A_3x_3 = 0$$
$$B_1x_1 + B_2x_2 + B_3x_3 = 0 \ ?$$

2.7 Prove that the set S defined in Example 2.5 satisfies P1–P4.

2.8   Let v denote a nonzero vector in $\mathbf{R}^3$, and think of it as a column vector, or 3×1 matrix, with coordinates v_1, v_2, and v_3. Now v represents a point in $\mathbf{P}^2(\mathbf{R})$ with homogeneous coordinates (v_1, v_2, v_3).

(a) Given a 3×3 matrix A with real coefficients $a_{11}, \ldots, a_{ij}, \ldots, a_{33}$, and non-zero determinant, show that the linear transformation of $\mathbf{R}^3$, $v \mapsto Av$, determines a mapping $T_A : \mathbf{P}^2(\mathbf{R}) \to \mathbf{P}^2(\mathbf{R})$ of points given by

$$(v_1, v_2, v_3) \mapsto \left(\sum_{i=1}^{3} a_{1i}v_i, \sum_{i=1}^{3} a_{2i}v_i, \sum_{i=1}^{3} a_{3i}v_i \right).$$

(b) Show that if $\lambda \in \mathbf{R} \setminus \{0\}$ then $T_A = T_{\lambda A}$.

(c) Prove that T_A is an automorphism of the projective plane $\mathbf{P}^2(\mathbf{R})$.

Hint: If $\mathfrak{L}$ denotes the solution set of $c_1 x_1 + c_2 x_2 + c_3 x_3 = 0$, show that $T_A(\mathfrak{L})$ is the solution set of $\sum_{i=1}^{3} c_i' x_i' = 0$ where $c_i' = \sum_{j=1}^{3} b_{ji} c_j$ and $A^{-1} = (b_{ij})$. In other words, T_A transforms points by $x \mapsto Ax$ and lines by $c \mapsto (A^{-1})^T c$, where X^T denotes the transpose of a matrix X.

2.9   Refer to Exercise 2.2. Does every automorphism of $\mathbf{S}$ that sends points of ℓ into itself restrict to an automorphism of the affine plane $\mathbf{S}_0$? Conversely, does every automorphism of $\mathbf{S}_0$ extend to one of the projective plane $\mathbf{S}$?

2.10   Let π be a projective plane of order n, i.e. it has $N = n^2 + n + 1$ points.

(a) Show that π has N lines.

Label the points $P_1, \ldots, P_N$ and lines $\ell_1, \ldots, \ell_N$. Let $a_{ij} = 1$ if point P_i is incident with ℓ_j, and $a_{ij} = 0$ if $P_i \not\in \ell_j$. This will define an $N \times N$ matrix $A = (a_{ij})$ of 0's and 1's called the *incidence matrix* of π.

(b) Show that $B = AA^T$ is an $N \times N$ matrix with each diagonal element $b_{ii} = n + 1$ and each off-diagonal element $b_{ij} = 1$.

(c) Show the converse of (b): Let n be an integer ≥ 2, let $N = n^2 + n + 1$, and let $A = (a_{ij})$ be a $N \times N$ matrix with $a_{ij} \in \mathbf{Z}_+ \cup \{0\}$ for all $i, j = 1, \ldots, N$. If A satisfies the equations $AA^T = A^T A = B$ (where B is the matrix defined in (b), then A is a matrix consisting of 0's and 1's and the incidence matrix of a projective plane.

2.11   Which of the geometric figures below belong most naturally to Euclidean geometry? To affine geometry? To projective geometry?

(a) Square.

(b) Parallelogram.

(c) Triangle.

(d) Trapezoid.

(e) Quadrilateral.

(f) Circles.

(g) Hyperbolas.

(h) Ellipses.

(i) Right triangle.

2.12* In the Euclidean plane, consider two intersecting lines ℓ and ℓ' and a point O not on either line. Show that the central projection $X \in \ell \mapsto X' = \ell'.OX$ is a one-to-one correspondence if and only if ideal points are included.

Chapter 3

Desargues' Theorem and the Principle of Duality

3.1 The Axiom P5 of Desargues

The first main result of projective geometry that we shall study is the theorem of Desargues about triangles in perspective. A *triangle* in an abstract projective plane consists of three noncollinear points (and the three nonconcurrent lines determined by these) — Axiom P3 states the existence of a triangle. Consider the truth value of the following simple statement in an abstract projective plane.

> **P5. Desargues' axiom** Let ABC and $A'B'C'$ be two triangles such that the three lines joining corresponding vertices, namely AA', BB' and CC', meet at a point O. (We say the two triangles are perspective from O.) Then the three pairs of corresponding sides intersect in three points $P = AB.A'B'$, $Q = AC.A'C'$, and $R = BC.B'C'$ which lie on a line. (We say that triangles ABC and $A'B'C'$ are perspective from the line PQR.)

Now it would not be quite right for us to call this a theorem, because it cannot be proved from our axioms P1–P4. However, we will show that it is true in the real projective plane (and this is the content of "Desargues' theorem"). Then we will take this statement as a further axiom, P5, or Desargues' axiom, of our abstract projective geometry. We will show by an example that P5 is not a consequence of P1–P4: namely, we will exhibit a geometry that satisfies P1–P4 but not P5. We will sometimes refer to projective planes satisfying Axiom P5 as *Desarguesian*[1] planes.

---

[1] Day-sargz-ian

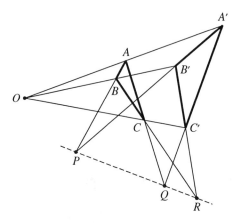

Figure 3.1. Desargues' axiom.

3.1. Desargues' theorem *In the real projective plane, two triangles perspective from a point are perspective from a line.*

Proof A straightforward computation using homogeneous coordinates follows. Let triangles ABC and $A'B'C'$ be perspective from the point O, and define the points $P = AB.A'B'$, $Q = AC.A'C'$ and $R = BC.B'C'$ (see Figure 3.1). We must show that P, Q and R are collinear.

Note that no three points of A, B, C and O are collinear; in other words, the coordinates of three of these points will form a linearly independent set of vectors in $\mathbf{R}^3$, while the fourth point has coordinates a linear combination of these with no coefficient zero. A simple linear change of coordinates followed by a scaling on the axes allows us with no loss of generality to assume $A = (1,0,0)$, $B = (0,1,0)$, $C = (0,0,1)$ and $O = (1,1,1)$. (This can equally well be said using an automorphism of the real projective plane: cf. Exercise 2.8.)

Now $A' \in OA$ which has the parametric equation

$$\lambda(1,0,0) + \mu(1,1,1) = (\lambda + \mu, \mu, \mu)$$

which is equivalent to $(1 + \frac{\lambda}{\mu}, 1, 1)$ or $(a', 1, 1)$ in homogeneous coordinates. Similarly, $B' = (1, b', 1)$ and $C' = (1, 1, c')$ for some $b', c' \in \mathbf{R}$.

Now we determine coordinates for P, Q and R. The line AB has equation $x_3 = 0$, while $A'B'$ has equation

$$\begin{vmatrix} x_1 & x_2 & x_3 \\ a' & 1 & 1 \\ 1 & b' & 1 \end{vmatrix} = (1 - b')x_1 + (1 - a')x_2 + (a'b' - 1)x_3 = 0.$$

P has homogeneous coordinates satisfying both equations; namely,

$$P = (1 - a', b' - 1, 0).$$

Similarly, we compute AC: $x_2 = 0$, and $A'C'$: $(c'-1)x_1 + (1-a'c')x_2 + (a'-1)x_3 = 0$ with point of intersection

$$Q = (a'-1, 0, 1-c').$$

Finally, BC: $x_1 = 0$, and $B'C'$: $(b'c'-1)x_1 + (1-c')x_2 + (1-b')x_3 = 0$ so

$$R = (0, 1-b', c'-1).$$

We conclude by noting that P, Q, and R are collinear, since the three representative vectors P, Q, and R form a linearly dependent set: using the particular coordinates given above, $P + Q + R = (0,0,0)$. Hence P, Q, and R lie on a plane through the origin.   □

A second, synthetic proof of this theorem is offered in Theorem 3.6.

3.2. Definition A *configuration* is a set, whose elements are called *points*, and a collection of subsets, called *lines*, which satisfies the following axiom:

C1. Two distinct points lie on at most one line.

It follows that two distinct lines have at most one point in common. Note however that two points may have no line joining them. Projective planes, affine planes, and the next example are configurations.

3.3. Example *Desargues' configuration* consists of 10 points and 10 lines, where each point lies on 3 lines, and each line contains 3 points. It is usually given the symbol 10_3.

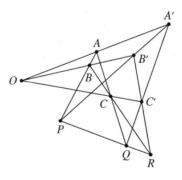

Figure 3.2. The configuration 10_3.

This configuration has a lot of symmetry. The role of the various points is not fixed Any one of the 10 points can be taken as the center of perspectivity of two triangles (Exercise 3.8). In Exercises 4.11–4.13, you will see that the "automorphism group" is the full group of permutations on five letters, which is related to viewing the figure below spatially: note the five planes.

3.2   Moulton's Example

We now give an example of a non-Desarguesian projective plane, that is, a plane satisfying P1–P4, but not P5. This will show that P5 is not a logical consequence of P1–P4.

A very simple idea for making Desargues' axiom fail in a projective plane is to let the line QR in Desargues' configuration (Figure 3.1) veer away from P; and this simple idea can be made to work! At the level of axioms P1–P4, lines are still, with few restrictions, the things we define them to be, and need not look much like the "shortest path between points."

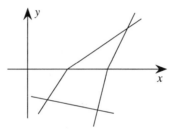

Figure 3.3. A Moulton triangle.

We define on $\mathbf{R}^2$ an alternative affine plane $\mathbf{A}'$, the *Moulton plane*. Points, vertical lines, and lines of negative slope are the same in $\mathbf{A}'$ as in the Euclidean affine plane. However, lines of positive slope are not admitted in $\mathbf{A}'$. Rather a line of slope $m > 0$ in the lower half-plane is pasted together at the x-axis with a line of slope $m/2$ in the upper half-plane. Analytically these *Moulton lines* are given by

$$f(x) = \begin{cases} m(x - x_0) & \text{if } x \le x_0 \\ \frac{m}{2}(x - x_0) & \text{if } x > x_0 \end{cases} \qquad (x_0 \in \mathbf{R}).$$

It is easy to see that two points in $\mathbf{A}'$ are joined by lines if they give a negative slope or lie vertically; in Exercise 3.1 you will be asked to check the existence of a Moulton line if the two points give a positive slope. Axiom A2 is verified by taking any line ℓ of $\mathbf{A}'$, a point P off ℓ, and drawing a line m through P with the same slope as ℓ in the upper and lower halfplanes: so, clearly $m \parallel \ell$ in $\mathbf{A}'$. A3 is a triviality.

Now complete $\mathbf{A}'$ to $\mathbf{P}'$, a projective plane (cf. Proposition 2.2). Arrange Desargues' configuration in $\mathbf{R}^2$ so that all points but P lie below the x-axis and so that

QR has positive slope. By the ordinary Desargues' theorem the Moulton line QR does not contain P.

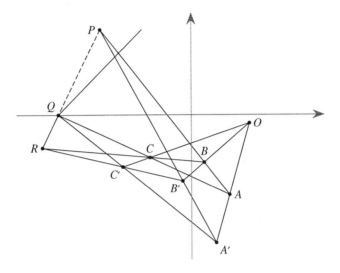

Figure 3.4. The Desargues configuration in the Moulton plane.

3.3   Axioms for Projective Space

In this section, we will obtain a theorem that will be needed in Chapter 8, where our coordinates no longer satisfy the commutative law of multiplication.

3.4. Definition A *projective 3-space* is a set whose elements are called *points*, together with certain subsets called *lines*, and certain other subsets called *planes*, which satisfy the following axioms:

> **S1.** Two distinct points P, Q lie on one and only one line ℓ.
> **S2.** Three noncollinear points P, Q, R lie on a unique plane.
> **S3.** A line meets a plane in at least one point.
> **S4.** Two planes have at least a line in common.
> **S5.** There exist four noncoplanar points, no three of which are collinear.
> **S6.** Every line has at least three points.

3.5. Example By a process analogous to that of completing an affine plane to a projective plane, the ordinary Euclidean 3-space can be completed to a projective 3-space, which we call *real projective 3-space* (Exercise 3.6). Alternatively, this same real projective 3-space can be described by homogeneous coordinates, as follows. A point is described by a quadruple (x_1, x_2, x_3, x_4) of real numbers, not all zero, where we agree that (x_1, x_2, x_3, x_4) and $(\lambda x_1, \lambda x_2, \lambda x_3, \lambda x_4)$ represent the same point, for any $\lambda \in \mathbf{R} \setminus \{0\}$. A plane is defined by a linear equation $\sum_{i=1}^{4} a_i x_i = 0$, not all $a_i = 0$, $a_i \in \mathbf{R}$, and a line is defined as the intersection of two distinct planes. The details of verification of the axioms are left to the reader in Exercise 3.7, and the reader should also check that the lines and points contained in the plane $x_1 = 0$ form the real projective plane defined in Section 2.2.

The remarkable fact is that although P5 is not a consequence of P1–P4 in the projective plane, it is a consequence of the seemingly equally simple axioms for projective 3-space. In Exercise 3.5 you will be asked to prove that a plane in projective 3-space is a projective plane. As a consequence the next theorem provides a second proof that Desargues' axiom is true in the real projective plane.

3.6. Theorem *Desargues' axiom holds in any projective 3-space, where we do not necessarily assume that all the points lie in a plane. In particular, Desargues' axiom holds for any plane that lies in a projective 3-space.*

> **Proof** We work with the points in Figure 3.1. There are two cases to consider.
>
> Case 1. Let us assume that the plane Σ containing the points A, B, C is different from the plane Σ' containing the points A', B', C'. The lines AB and $A'B'$ both lie in the plane determined by O, A, B, and so they meet in a point P. Similarly we see that AC and $A'C'$ meet, and that BC and $B'C'$ meet. Now the points P, Q, R lie in the plane Σ, and also in the plane Σ'. Hence they lie in the intersection $\Sigma \cap \Sigma'$, which is a line (Exercise 3.2c).

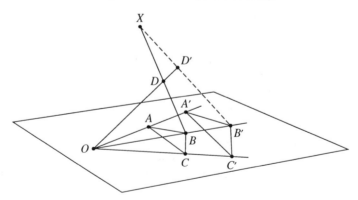

Figure 3.5. Desargues configuration embedded in projective 3-space.

> Case 2. Suppose that $\Sigma = \Sigma'$, so that the whole configuration lies in one plane Σ. Let X be a point which does not lie in Σ. One has lines joining X to

all the points in the diagram. Let D be a point on XB, different from B, and let $D' = OD.XB'$. (Why do OD and XB' meet?) Then the triangles ADC and $A'D'C'$ are perspective from O, and do not lie in the same plane. We conclude from Case 1 that the points $P' = AD.A'D'$, $Q = AC.A'C'$, and $R' = DC.D'C'$ lie in a line. But these points are projected from X onto P, Q, and R, hence P, Q, R are collinear. $\qquad\qquad\qquad\qquad\qquad\qquad\qquad\qquad\qquad\qquad$ □

3.4 Principle of Duality

We now proceed to cut in half our labors by giving the principle of duality. Before reading the next proposition it will be helpful to recall that a projective plane is a set where elements are called "points" together with a set of subsets of those "points", each of which is called a "line": the "points" and "lines", though, must satisfy P1–P4. Instead of using quotation marks, we write point and line in different script from time to time in our discussion of duality.

3.7. Proposition *Let π be a projective plane. Let π^* be the set of lines in π, and define a line in π^* to be a pencil of lines in π. Then π^* is a projective plane. Furthermore, if π satisfies P5, so does π^*.*

3.8. Remark We call π^* the *dual projective plane* of π.

> **Proof** We must verify the axioms P1–P4 for π^*. These translate into statements D1, D2, D3, and D4, respectively, which we show to be simple consequences of P1–P4. We also show P5 $\Longrightarrow$ D5.
> P1. If p, q are two distinct *points* of π^*, then there is a unique *line* of π^* containing p and q. If we translate this into a statement for π, it says
>
> > **D1.** If p, q are two distinct lines of π, then there is a unique pencil of lines containing p, q.
>
> I.e. p, q have a unique point in common. Thus D1 is equivalent to P2.
> P2. If L and M are two *lines* in π^*, they have exactly one *point* in common. In π, this says that
>
> > **D2.** Two pencils of lines have exactly one line in common.
>
> This is equivalent to P1.
> P3. There are three noncollinear *points* in π^*.
>
> > **D3.** There are three nonconcurrent lines in π.
>
> (We say three or more lines are *concurrent* if they all pass through some point, i.e. if they are contained in a pencil of lines.) By P3 there are three noncol-

linear points A, B, C. Then one sees easily that the lines AB, AC, BC are not concurrent: these correspond to three noncollinear *points* in π^*. (Conversely, three nonconcurrent lines implies the existence of three noncollinear points.)

P4. Every *line* in π^* has at least three *points*. This says that

D4. Every pencil in π has at least three lines.

Let the pencil be centered at P, and let ℓ be some line not passing through P. Then by P4, ℓ has at least three points A, B, C. Hence the pencil of lines through P has at least three lines $a = PA$, $b = PB$, $c = PC$. (Conversely, assuming D4 we easily show P4.)

Now we will assume P5, Desargues' axiom, is true in π and prove it in π^*.

P5. Let o, a, b, c, a', b', c' be seven distinct *points* of π^*, such that oaa', obb', occ' are collinear, and abc, $a'b'c'$ are triangles. Then the *points* $p = ab.a'b'$, $q = ac.a'c'$, and $p = bc.b'c'$ are collinear.

Translated into π, this says the following

D5. Let o, a, b, c, a', b', c' be seven lines, such that o, a, a'; o, b, b'; o, c, c' are concurrent and such that abc and $a'b'c'$ form two triangles. Then the lines $p = (a.b) \cup (a'.b')$, $q = (a.c) \cup (a'.c')$, and $r = (b.c) \cup (b'.c')$ are concurrent.

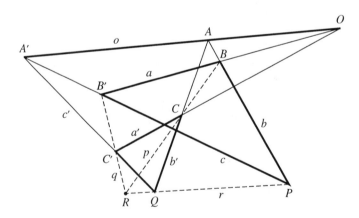

Figure 3.6. Proof of the converse of Desargues' axiom.

To prove this statement, we will label the points of the diagram in such a way as to be able to apply P5. So let $O = o.a.a'$, $A = o.b.b'$, $A' = o.c.c'$, $B = a.b$, $B' = a.c$, $C = a'.b'$, and $C' = a'.c'$. Then O, A, B, C, A', B', C' satisfy the hypothesis of P5, so we conclude that $P = AB.A'B' = b.c$, $Q = AC.A'C' = $

$b'.c'$, and $R = BC.B'C' = p.q$ are collinear. But $PQ = r$, so p, q, and r are concurrent at R.   □

We have in fact nearly proven that π is a projective plane if and only if π^* is a projective plane; indeed, π is Desarguesian iff π^* is Desarguesian (Exercise 3.11).

3.9. The converse of Desargues' axiom Notice that statement D5 is in fact the converse of Desargues' axiom: two axially perspective triangles are centrally perspective. What we have proved then is that if Desargues' axiom holds in a projective plane, then so does its converse.[2]

3.10. Principle of duality *Let S be any statement about a projective plane π, which can be proved from the axioms P1–P4 (respectively P1–P5). Then the "dual" statement D, obtained from S by interchanging the words*

$$point \longleftrightarrow line$$
$$lies\ on \longleftrightarrow passes\ through$$
$$collinear \longleftrightarrow concurrent$$
$$intersection \longleftrightarrow join$$
$$etc.$$

is a true statement as well about projective planes (resp. Desarguesian planes).

Metaproof The statement S is true for projective planes, so it holds in particular for the dual projective planes, which are themselves projective planes as we have just seen. Now if S is true for π^*, then D is true for π, since D is just the application to S of the definition of *point, line, point on line*, etc. in π^*. π being arbitrary, D is a true statement about projective planes.   □

The next two propositions provide an example of the principle of duality in application.

3.11. Proposition *Suppose a projective plane π contains a line ℓ with $n+1$ points. Then there are in all $n^2 + n + 1$ points in π.*

Proof By P3, there is a point $P \notin \ell$. Now every other point Q determines a line PQ which intersects ℓ at the point $PQ.\ell$. This being a bijection between the points of ℓ and the pencil of lines at P, there are $n+1$ lines in the pencil

---

[2]That the dual of statement P5 is its converse is by no means usual for statements about projective planes.

at P. Counting first $n + 1$ points on PQ, then only n new points on a second line through P, we arrive at

$$n + 1 + n \cdot n = n^2 + n + 1$$

points in π. $\square$

The proposition is dualized as follows.

3.12. Proposition *Suppose a projective plane π contains a point P with a pencil of $n + 1$ lines. Then there are in all $n^2 + n + 1$ lines in π.*

The proof, though ordinarily omitted, is simply an application of Proposition 3.11 to π^*. Since pencils of lines and *ranges* of points (i.e., points on a line) are in bijective correspondence, it follows from the two propositions that there are as many lines as there are points in a finite projective plane.

3.13. Remark Consider the dual of the dual projective plane π^*; denote it by π^{**}. Is it something new? There is a natural map $\pi \to \pi^{**}$ given by sending a point P of π into the pencil of lines through P, which is a point of π^{**}. This turns out to be an isomorphism of the two projective planes π and π^{**} (Exercise 3.9). Hence π^{**} is not anything new.

For the real projective plane π the dual projective plane π^* is isomorphic with π (in notation, $\pi \cong \pi^*$). Given a line ℓ through the origin O, send this to the plane through O perpendicular to ℓ: check that this defines an isomorphism. It is also true that $\pi \cong \pi^*$ for the finite projective planes π of $p^{2n} + p^n + 1$ points (p a prime, n a positive integer). Isomorphisms between π and π^* are called *polarities* in the literature. However, there are non-Desarguesian projective planes where $\pi \not\cong \pi^*$.

Exercises

3.1   Establish the existence of a Moulton line through two points giving positive slope.

3.2   Using the axioms S1–S6 of projective 3-space, prove the following statements. Be very careful not to assume anything except what is stated by the axioms. Refer to the axioms explicitly by number.

      (*a*) If two distinct points P, Q lie in a plane Σ, then the line joining them is contained in Σ.

      (*b*) A plane and a line not contained in the plane meet in exactly one point.

      (*c*) Two distinct planes meet in exactly one line.

      (*d*) A line and a point not on it lie in a unique plane.

3.3   Provide the details of case 2 of the proof of Theorem 3.6 by drawing on the axioms and Exercise 3.2.

3.4   Given two lines ℓ and m intersecting "off the paper", and a point P not on either line, use Desargues' theorem to construct a line through P and $\ell \cap m$.

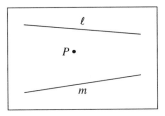

Figure 3.7. Two lines intersecting "off the paper".

3.5   Prove that any plane Σ in a projective 3-space is a projective plane, i.e. satisfies the axioms P1–P4. (You may use the results of Exercise 3.2).

3.6   Propose axioms for affine 3-space and show how one may complete this to obtain a projective 3-space. Check carefully that Axioms S1–S6 are satisfied. State and prove a 3-dimensional analog of Proposition 2.9.

3.7   Verify that the projective space $\mathbf{P}^3(\mathbf{R})$ does indeed satisfy Axioms S1–S6.

3.8   Find the two triangles in the configuration 10_3 (see Figure 3.2), which are in perspective from center P. What is their *axis* (i.e., the line in which their corresponding sides meet)?

3.9   Show that the map $T: \pi \rightarrow \pi^{**}$ defined by $T(P) = [P]$ (= pencil of lines through P) is an isomorphism of projective planes.

3.10   Convince yourself that statements D1–D4 in Proposition 3.7 are an alternative and equivalent set of axioms for a projective plane.

3.11   Show that the converse of Desargues' axiom implies Desargues' axiom.

3.12   Find the finite projective space of least number of points. List the points, lines and planes.

Chapter 4

A Brief Introduction to Groups

The sets of automorphisms with their compositions, Tran **A** and Dil **A**, which we encountered when studying the affine plane **A**, are examples of an important algebraic concept, the group. Groups got their real start with Galois as finite invariants of polynomial equations, which could resolve questions of solvability by radicals. In Galois theory, difficult questions about polynomial equations are converted to easier questions about groups, which can be answered with a small amount of group theory. Group theory and projective geometry have an even more fundamental relationship, reflected in the quotes below.

> "Whenever you have to do with a structure-endowed entity Σ, try to determine its group of automorphisms... You can expect to gain a deep insight into the constitution of Σ in this way."
>
> Hermann Weyl (1885–1955)
> ("Symmetry")

> "A geometry is defined by a group of transformations, and investigates everything that is invariant under the transformations of this given group."
>
> Felix Klein (1849-1925)

4.1   Elements of Group Theory

4.1. Definition A *group* is a set G, together with a function (or binary operation) $G \times G \to G$, $(a,b) \mapsto ab$, such that

35

G1. *Associativity:* $(ab)c = a(bc)$ for all $a, b, c \in G$.
G2. There exists an element $1 \in G$ such that $a \cdot 1 = 1 \cdot a = a$ for all a.
G3. To each $a \in G$, there exists an element $a^{-1} \in G$ such that $aa^{-1} = a^{-1}a = 1$.

In writing the value of the binary operation on (a,b) simply as ab, we have supressed the function notation, such as $f(a,b)$. The element 1 is called the *identity, unit,* or *neutral element.* The element a^{-1} is called the *inverse* of a. (Both 1 and a^{-1} are easily shown to be unique.) The cancellation law holds in groups: $ax = ay$ or $xa = ya$ implies $x = y$, where $a, x, y \in G$. A set with binary operation satisfying only G1 is called a *semigroup.*

In general, the product ab may be different from ba. However, we say the group G is *abelian,* or *commutative,* if

G4. $ab = ba$ for all $a, b \in G$.

The multiplicative notation for a group is often changed to an additive notation for abelian groups: $a + b$, neutral element 0, inverse $-a$ of a. E.g., the common integers **Z** under addition is an abelian group. Another example is a vector space; forgetting all about scalar multiplication, one notes that the vectors form an abelian group under addition. The next two examples are almost never abelian.

4.2. Example Let S be any set, and let $G = \mathrm{Perm}\, S$ be the set of permutations of the set S. (Recall that a permutation is a one-to-one mapping of S onto S.) If $g_1, g_2 \in G$ are two permutations, define $g_1 g_2 \in G$ to be the permutation obtained by performing first g_2, then g_1, i.e., $g_1 g_2(x) = g_1(g_2(x))$ for all $x \in S$.

If S is a set with 3 or more elements, then $\mathrm{Perm}\, S$ is a nonabelian group.

4.3. Example Let C be a configuration, and let G be the set of automorphisms of C, i.e. the set of those permutations of C *which send lines onto lines.* Again we define the product $g_1 g_2$ of two automorphisms g_1, g_2, by performing g_2 first, and then g_1. This group is written $\mathrm{Aut}\, C$. Since affine planes and projective planes are configurations, we now have a definition of their automorphism groups.

4.4. Definition A *homomorphism* $\phi \colon G_1 \to G_2$ of one group to another, is a mapping of the set G_1 to the set G_2 such that

$$\phi(ab) = \phi(a)\phi(b)$$

for each $a, b \in G_1$.

An *isomorphism* of one group with another, is a homomorphism which is one-to-one and onto. It follows easily that a homomorphism sends unit to unit. Linear transformations of vector spaces provide examples of homomorphisms between (abelian) groups. An example of an isomorphism is $\phi(x) = e^x$, which provides an isomorphism between the additive group of reals $\langle \mathbf{R}, +, 0 \rangle$ and the multiplicative group of positive reals $\langle \mathbf{R}_+, \cdot, 1 \rangle$.

4.5. Definition Let G be a group. A *subgroup* of G is a nonempty subset $H \subseteq G$, such that for any $a, b \in H$, $ab \in H$, and $a^{-1} \in H$.

 Note this condition implies $1 \in H$. Then H, with the binary operation restricted to H, is itself a group.

4.6. Example Let G be the group of permutations of a set S. Let $x \in S$, and let $H_x = \{ g \in G \mid g(x) = x \}$. Then H_x is a subgroup of G, called the *stabilizer subgroup* of x (Exercise 4.7).

4.7. Example The even integers under addition form a subgroup of $\mathbf{Z}$. More generally, the multiples of n form a subgroup $n\mathbf{Z}$ under addition.

4.8. Example $\mathrm{Aut}\, C$ is a subgroup of $\mathrm{Perm}\, C$, where C denotes both a configuration and its underlying set of points.

4.9. Example If $\mathbf{A}$ denotes an affine plane, $\mathrm{Tran}\, \mathbf{A}$ is a subgroup of $\mathrm{Dil}\, \mathbf{A}$ which is in turn a subgroup of $\mathrm{Aut}\, \mathbf{A}$ by the results of Section 1.2.

4.10. Definition Let G be a group, and H a subgroup of G. The *left coset* of H generated by $g \in G$ is the subset of G given by $gH = \{ gh \mid h \in H \}$. The *right coset* Hg is similarly defined to be the subset $\{ hg \mid h \in H \}$.

4.11. Lemma *Let H be a subgroup of G, and let gH be a left coset. Then there is a one-to-one correspondence between the elements of H and the elements of gH. (In particular, if H is finite, they have the same number of elements.)*

 Proof Map $H \to gH$ by $h \mapsto gh$. By definition of gH, this map is onto. So suppose $h_1, h_2 \in H$ have the same image. Then $gh_1 = gh_2$. Multiplying on the left by g^{-1}, we deduce $h_1 = h_2$. $\square$

4.12. Lagrange's theorem *Let G be a finite group, and let H be a subgroup. Then*

$$\#(G) = \#(H) \cdot (\text{number of left cosets of } H).$$

 Proof Indeed, all the left cosets of H have the same number of elements as H, by the lemma. If $g \in G$, then $g \in gH$, since $g = g \cdot 1$, and $1 \in H$. Thus G is the union of the left cosets of H. Finally, if we show that two cosets gH, and $g'H$ are either equal, or disjoint, then we have the set G partitioned into subsets of size $\#(H)$ each, which proves the theorem.

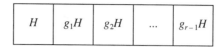

Figure 4.1. Partition of G into r equal sized left cosets of H.

Indeed, suppose gH and $g'H$ have an element in common, namely x. Then $xH \subseteq gH$ and $xH \subseteq g'H$, but the lemma tells us that cosets have the same finite cardinality. Hence $gH = xH = g'H$.   □

Thus the *order* of the subgroup H, i.e. the number of elements in H, divides the order of G.[1] Let us see a good application of Lagrange's theorem: we take the first step towards classification of finite groups up to isomorphism.

In a group G, an element g generates a subgroup $\langle g \rangle$ consisting of the powers of g: $g^2 = gg$, $g^3 = g^2g$, etc. (includes also g^{-1}, $g^{-2} = g^{-1}g^{-1}$, if necessary). In a finite group, the order of $\langle g \rangle$ is called the *order of g*, and equals the least positive integer n such that $g^n = 1$.

If G has prime order, Lagrange's theorem informs us that there are no non-trivial subgroups (different from $\{1\}$ and G), so every element $\neq 1$ generates all of G. Groups generated by one element are called *cyclic groups*, and any two cyclic groups of the same order are isomorphic (Exercise 4.9). Thus, groups of prime order are cyclic, and any two groups of same prime order are isomorphic.

4.13. Definition Suppose G is a subgroup of some Perm S. The *orbit* of x is the set of points
$$\beta_x = \{g(x) \mid g \in G\}$$

4.14. Corollary *If G is a finite subgroup of* Perm S, *then*
$$\#(G) = \#(H_x) \cdot \#(\beta_x).$$

Proof In order to apply Lagrange's theorem, we need only show the orbit β_x in one-to-one correspondence with the number of distinct left cosets of H_x in G.

Given $y \in \beta_x$, there exists $g \in G$ such that $y = g(x)$, so map
$$\beta_x \to \{\text{left cosets}\}$$

with the function $T : y \mapsto gH_x$.

---

[1]Beware though that it is not necessarily true that every divisor of $\#(G)$ corresponds to $\#(H)$ for some subgroup H in G: there is no six element subgroup in the 24 element group S_4 whose acquaintance we make below.

We claim T is a one-to-one correspondence. T is well-defined: if $g'(x) = y = g(x)$, then $g^{-1}g'(x) = x$, so $g^{-1}g' \in H_x$, hence $g'H_x = gH_x$, i.e. T is single-valued at y. T is onto: let gH_x be a coset, then $T(g(x)) = gH_x$. T is one-to-one: if $T(y_1) = gH_x = T(y_2)$, then $y_1 = g(x) = y_2$. Hence, T is the desired one-to-one correspondence.   □

4.15. Definition A group $G \subseteq \operatorname{Perm} S$ of permutations of a set S is *transitive* if the orbit of some element is all of S (in that case, the orbit of any element is all of S). If G is transitive, $\#(G) = \#(H_x) \cdot \#(S)$, by using the last corollary.

4.16. Corollary *Let S be a set with n elements, and let $S_n = \operatorname{Perm} S$. Then*

$$\#(S_n) = n!.$$

Proof by induction on n If $n = 1$, there is only the identity permutation, so $\#(S_1) = 1$. Induction hypothesis: Assume that $\#(S_m) = m!$ for $m \le n$. Let S have $n + 1$ elements, and let $x \in S$. Let H_x be the subgroup of permutations leaving x fixed. S_{n+1} is transitive, since one can permute x with any other element of S. Hence

$$\#(S_{n+1}) = \#(S) \cdot \#(H_x) = (n + 1) \cdot \#(H_x).$$

But H_x is just the group of permutations of the remaining n elements of S different from x, so $\#(H_x) = n!$ by the induction hypothesis. Hence $\#(S_{n+1}) = (n + 1)!$ □

4.17. Generators A subset A of a group G is said to *generate* the subgroup H, denoted by $H = \langle A \rangle$, if H is the smallest subgroup containing A. H is in fact the subgroup

$$\{a_1^{n_1} \dots a_q^{n_q} \mid a_1, \dots, a_q \in A; \, n_1, \dots, n_q \in \mathbf{Z}\}$$

of products of powers of elements in A (Exercise 4.16).

Later in the course, we will have much to do with the group of automorphisms of a projective plane, and certain of its subgroups. In particular, we will show that Desargues' axiom P5 is equivalent to the statement that the group of automorphisms has enough (of the projective equivalents of) translations and stretchings. In Section 4.2 and the exercises, we will content ourselves with calculating the automorphisms of a few simple configurations.

4.2   Automorphisms of the Projective Plane of 7 Points

Let π be the projective plane of 7 points A, B, C, D, P, Q, R. π may be obtained by completing the affine plane of four points A, B, C, D. Its lines are indicated in the figure below.

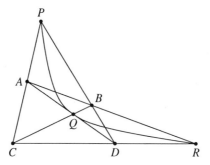

Figure 4.2. A pictorial representation of the seven point projective plane π.

4.18. Proposition $G = \operatorname{Aut} \pi$ *is transitive.*

Proof We will write down some elements of G explicitly.

$$a = (AC)(BD)$$

for example. This notation means "interchange A and C, and interchange B and D". More generally a symbol $(A_1 A_2 \ldots A_r)$ means "send A_1 into A_2, A_2 into A_3, $\ldots$, A_{r-1} into A_r, and A_r into A_1", and is referred to as an r-*cycle*. Multiplication of two such symbols is defined by performing the one on the right first, then the next from right to left and so on.

$$b = (AB)(CD)$$

Thus we see already that A can be sent to B or to C. We calculate

$$ab = (AC)(BD)(AB)(CD) = (AD)(BC),$$

and

$$ba = (AB)(CD)(AC)(BD) = (AD)(BC) = ab.$$

Thus we can also send A to D.

Another automorphism is

$$c = (BQ)(DR).$$

Since the orbit of A already contains B, C, D, we see that it also contains Q and R. Finally

$$d = (PA)(BQ)$$

shows that the orbit of A is all of π, so G is transitive. $\square$

4.19. Lemma *Let $H_P \subseteq G$ be the subgroup of automorphisms of π leaving P fixed. Then H_P is transitive on the set $\pi \smallsetminus \{P\}$.*

Proof Note that a, b, c above are all in H_P, so that the orbit of A under H_P is

$$\{A, B, C, D, Q, R\} = \pi \smallsetminus \{P\}.$$

$\square$

4.20. Theorem *G has 168 elements.*

Proof We carry the above analysis a step farther as follows. Let $K \subseteq H_P$ be the subgroup leaving Q fixed. Therefore, since elements of K leave P and Q fixed, they also leave R fixed. K is transitive on the set $\{A, B, C, D\}$, since $a, b \in K$. On the other hand, an element of K is uniquely determined by where it sends the point A, since lines go to lines and two of the three points per line are determined. Since A may only go to four points, K is just the group consisting of the four elements $1, a, b, ab$. We conclude from the previous discussion that

$$\#(G) = \#(H_P) \cdot \#(\pi) = 7 \cdot \#(H_P)$$

$$\#(H_P) = \#(K) \cdot \#(\pi \smallsetminus \{P\}) = 4 \cdot 6$$

whence $\#(G) = 7 \cdot 6 \cdot 4 = 168$. $\square$

4.21. Corollary *Given triangles $A_1 A_2 A_3$ and $A_1' A_2' A_3'$ in π, there is a unique automorphism sending $A_i \mapsto A_i'$ $(i = 1, 2, 3)$.*

Proof For each triangle $\triangle B_1 B_2 B_3$ we show that there is an automorphism $\alpha_{B_1 B_2 B_3}$ sending P, Q, A onto B_1, B_2, B_3, respectively. Then

$$\alpha_{A_1' A_2' A_3'} \circ \alpha_{A_1 A_2 A_3}^{-1}$$

proves the existence part of our statement.

Since G is transitive, we can find $g \in G$ such that $g(P) = B_1$. Since $B_1 \neq B_2$ it follows that $g^{-1}(B_2) \neq P$. But H_P is transitive on $\pi \smallsetminus P$, so there is an element $h \in H_P$ such that $h(Q) = g^{-1}(B_2)$. Then $gh(P) = B_1$ and $gh(Q) = B_2$. Now $(gh)^{-1}(B_3) \notin \{P, Q, R\}$ since B_3 is not on the line $B_1 B_2$. Hence there is $k \in K$ such that $k(A) = (gh)^{-1}(B_3)$. Then $(ghk)(P) = B_1$, $(ghk)(Q) = B_2$, and $(ghk)(A) = B_3$. This completes the existence argument.

For uniqueness of this element, let us count the number of ordered triples of noncollinear points in π. The first can be chosen in 7 ways, the second in 6 ways, and the last in 4 ways. Thus there are 168 such triples. Since the order of G is 168, there must be exactly one automorphism sending a given triangle into another triangle. $\square$

Exercises

4.1   Show that the identity of a group G is unique. Next, show that inverses are unique: if b and c satisfy $ab = ba = 1$, $ac = ca = 1$, then $b = c$.

4.2   Given a semigroup G, a *left identity* is an element e satisfying $ea = a$ (for all $a \in G$). Given $a \in G$, a *left inverse* b satisfies $ba = e$. Show that a semigroup with left identity, in which each element has a left inverse, is a group. Is something similar true where right replaces left?

4.3   Let G be a group. If $a^2 = 1$ for all $a \in G$, prove that G is abelian.

4.4   Let n be a positive integer > 1. Consider the additive group $\mathbf{Z}$ of integers, and subgroup $n\mathbf{Z}$. Show that the left cosets $\{x + n\mathbf{Z} \mid x = 0, \dots, n-1\}$ form a group under $+$, call it $\mathbf{Z}_n$.

4.5   If $\phi\colon G \to H$ is a homomorphism, prove that the *kernel* $\mathrm{Ker}(\phi) = \{g \in G \mid \phi(g) = 1\}$ of ϕ is a subgroup of G.

4.6   Show that the natural map $\phi\colon \mathbf{Z} \to \mathbf{Z}_n$ given by $x \mapsto x + n\mathbf{Z}$ is a homomorphism with kernel $n\mathbf{Z}$.

4.7   If $G = \mathrm{Perm}\,S$, prove that $H_x = \{g \in G \mid g(x) = x\}$ is a subgroup of G for every $x \in S$. If $x = g(y)$ for some $g \in G$ and $x, y \in S$, show that $H_x = gH_yg^{-1}$; whence H_x and H_y are isomorphic.

4.8   Let $S = \{1, 2, 3\}$. Show that $\mathrm{Perm}\,S$ has two elements σ and τ for which $\sigma\tau \neq \tau\sigma$. (You might want to use the cycle notation suggested in the proof of Proposition 4.18.

4.9   Show that $\mathbf{Z}_n$ is a cyclic group, i.e. generated by one element. Prove that cyclic groups of same order n are isomorphic, whence all are isomorphic to $\mathbf{Z}_n$.

4.10   Prove in a manner similar to Section 4.2 that the affine plane of 9 points has automorphism group of order $9 \cdot 8 \cdot 6 = 432$, and any three noncollinear points can be taken into any three noncollinear points by a unique element of the group.

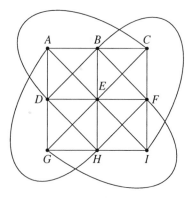

Figure 4.3. The affine plane of 9 points.

Exercises 4.11–4.13. We will consider the *Desargues configuration*, which is a set of 10 elements,

$$\Sigma = \{O, A, B, C, A', B', C', P, Q, R\},$$

and 10 lines, which are the subsets $\{O, A, A'\}$, $\{O, B, B'\}$, $\{O, C, C'\}$, $\{A, B, P\}$, $\{A', B', P\}$, $\{A, C, Q\}$, $\{A', C', Q\}$, $\{B, C, R\}$, $\{B', C', R\}$, and $\{P, Q, R\}$. In Exercises 4.11–4.13, let $G = \operatorname{Aut}\Sigma$ be the set of automorphisms of Σ.

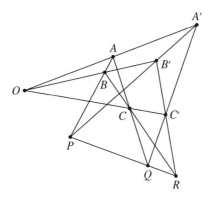

Figure 4.4. The Desargues configuration.

4.11   Show that G is transitive on Σ.

4.12   (*a*) Show that the subgroup of G leaving a point fixed is transitive on a set of six letters.

(*b*) Show that the subgroup of G leaving two collinear points fixed has order 2.

(*c*) Deduce the order of G from the previous results.

Now we consider some further subsets of Σ, which we will call planes, namely

$$1 = \{O, A, B, A', B', P\}$$
$$2 = \{O, A, C, A', C', Q\}$$
$$3 = \{O, B, C, B', C', R\}$$
$$4 = \{A, B, C, P, Q, R\}$$
$$5 = \{A', B', C', P, Q, R\}$$

4.13   Show that each element of G induces a permutation of the set of five planes

$$\{1, 2, 3, 4, 5\},$$

and that the resulting mapping $\phi\colon G \to \operatorname{Perm}\{1, 2, 3, 4, 5\}$ is an isomorphism of groups. Thus G is isomorphic to the permutation group on 5 letters.

4.14   Let S_4 be the group of permutations of the four symbols $1,2,3,4$.

(*a*) Let $G \subseteq S_4$ be the subgroup generated by the permutation (1234). What is the order of G?

(*b*) Let $H \subseteq S_4$ be the subgroup generated by the permutations (12) and (34). What is the order of H?

(*c*) Is there a group isomorphism $\phi \colon G \to H$? If so, write it explicitly. If not, explain why not.

4.15   The *Pappus configuration* Σ is the configuration of 9 points and 9 lines as shown in the diagram. Compute the order of the automorphism group of Σ.

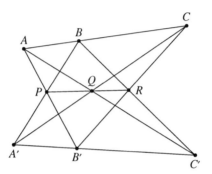

Figure 4.5. The Pappus configuration 9_3.

4.16   Let A be a subset of a group G.

(*a*) Show that the set $K = \{a_1^{n_1} \ldots a_q^{n_q} \mid a_1,\ldots,a_q \in A;\ n_1,\ldots,n_q \in \mathbf{Z}\}$ is a subgroup of G.

(*b*) Show that $K = \langle A \rangle$, the subgroup generated by A.

4.17   Show that the number of right cosets of a subgroup in a finite group is equal to the number of left cosets.

Chapter 5

Elementary Synthetic Projective Geometry

We will now define what we mean by central projection, or perspectivity, in our axiomatic development. We will also meet some of the basic objects of study in projective geometry, such as complete quadrangles, harmonic points, and the cross ratio of four collinear points. But first we must add another axiom to our projective geometry.

5.1 Fano's Axiom P6

Fano's axiom is a statement about the diagonal points of a complete quadrangle, which we define next.

5.1. Definition Suppose A, B, C, and D are four points in a projective plane such that no three of these points are collinear. Then the *complete quadrangle ABCD* is the collection of seven points and six lines obtained by taking all six lines determined by A, B, C, and D, and then taking the intersection of opposite sides: $P = AB.CD$, $Q = AC.BD$, and $R = AD.BC$. The points P, Q, R are called *diagonal points* of the complete quadrangle.

It may happen that the diagonal points P, Q, R of a complete quadrangle are collinear (as for example in the projective plane of seven points). However, this never happens in the real projective plane, as we will see below. In general, it is to be regarded as a somewhat pathological phenomenon, hence we will make an axiom saying this should not happen.

P6. Fano's axiom The diagonal points of a complete quadrangle are not collinear.

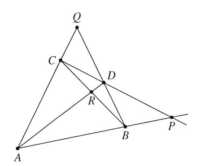

Figure 5.1. Complete quadrangle *ABCD*.

5.2. Proposition *The real projective plane satisfies P6.*

Proof Given a complete quadrangle *ABCD*, no three of A, B, C, D are collinear, so in the homogeneous coordinate model any three representative vectors (representing $A, B, \cdots =$ lines through the origin) are linearly independent. Choosing these as a basis and rescaling, we may assume A, B, C, D to be the points $(1,0,0)$, $(0,1,0)$, $(0,0,1)$, and $(1,1,1)$ respectively. You are asked to compute that the diagonal points of this complete quadrangle are $(1,1,0)$, $(1,0,1)$, and $(0,1,1)$ in Exercise 5.1*b*. To see if they are collinear, we compute the determinant:

$$\begin{vmatrix} 1 & 1 & 0 \\ 1 & 0 & 1 \\ 0 & 1 & 1 \end{vmatrix} = -2$$

Hence the rows are linearly independent, so the lines through the origin they represent are not coplanar, and we conclude that the points are not collinear. $\square$

Let's temporarily call a projective plane satisfying P6 a Fano plane. Now the projective plane of seven points shows that P1–P5 does not imply P6; i.e., Axiom P6 is independent of the previous axioms. In order to extend the principle of duality, we had better consider the question of whether the dual plane π^* of a Fano plane π is also a Fano plane. We will obviously need to dualize the notion of complete quadrangle.

5.3. Definition A *complete quadrilateral abcd* is the collection of seven lines and six points, obtained by taking four lines a, b, c, d, of which we require that no three are concurrent, their six points of intersection, and the three lines $p = (a.b) \cup (c.d)$, $q = (a.c) \cup (b.d)$, $r = (a.d) \cup (b.c)$ joining opposite pairs of points. The lines p, q, and r are called the *diagonal lines* of the complete quadrilateral.

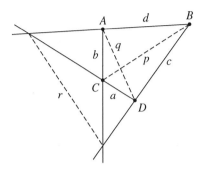

Figure 5.2. Complete quadrilateral *abcd*.

5.4. Proposition *If π is a Fano plane, then so is its dual plane π^*.*

Proof Now a complete quadrangle in π^* is a complete quadrilateral in π. Its diagonal points are diagonal lines in π. We must prove that the diagonal lines are necessarily nonconcurrent.

Suppose *abcd* is a complete quadrilateral in π. Its diagonal lines are p, q, and r as above. Let $A = b.d$, $B = d.c$, $C = a.b$, and $D = a.c$. If A, B, C are collinear, then $C \in AB = d$, i.e., a, b, and d are concurrent, which is a contradiction. Three similar arguments show that no three points of A, B, C, D are collinear.

Then consider the complete quadrangle $ABCD$. The diagonal points are $P = AB.CD = (b.d \cup d.c).(a.b \cup a.c) = d.a$, $Q = AC.BD = (b.d \cup a.b).(d.c \cup a.c) = b.c$, and $R = AD.BC = (b.d \cup a.c).(d.c \cup a.b) = q.p$. Note that $PQ = r$. Then $R \in PQ$ if and only if p, q, and r are concurrent. Hence, the diagonal points of $ABCD$ are collinear iff the diagonal lines of *abcd* are concurrent. $\quad\square$

5.5. Remark The reader is urged to construct the duals of definitions, theorems, and proofs. For example, in the next section try developing the theory of harmonic lines, dual to harmonic points.

5.2   Harmonic Points

5.6. Definition An ordered quadruple of distinct points A, B, C, D on a line is called a *harmonic quadruple* if there is a complete quadrangle $XYZW$ such that A and B are diagonal points of the complete quadrangle (say $A = XY.ZW$ and $B = XZ.YW$), and C, D lie on the remaining two sides of the quadrangle (say $C \in XW$ and $D \in YZ$).

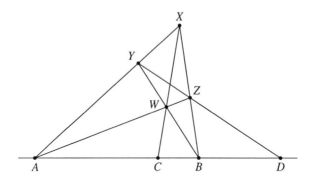

Figure 5.3. Harmonic points A, B, C, D.

In symbols, we write $H(A, B; C, D)$ if A, B, C, D form a harmonic quadruple. From the definition, we equally well have

$$H(B, A; C, D), \qquad H(A, B; D, C), \qquad H(B, A; D, C).$$

Note that if A, B, C, D is a harmonic quadruple, then the fact that A, B, C, D are distinct implies that the diagonal points of a defining quadrangle $XYZW$ are not collinear. In fact, the notion of four harmonic points does not make much sense unless Fano's axiom P6 is satisfied, hence we will always assume this axiom when we speak of harmonic points.

5.7. Proposition *Let A, B, C be three distinct points on a line. Then (assuming P6) there is a point D such that $H(A, B; C, D)$. Furthermore, if P5 is assumed, this point is unique. (D is called the* fourth harmonic point *of A, B, C, or the* harmonic conjugate *of C with respect to A and B.)*

> **Proof** We construct a complete quadrangle having A and B as diagonal points and with C on one of the two remaining lines.
>
> By D3 we find two lines ℓ and m through A, different from the line ABC. Find a line n through C, different from ABC.
>
> Then let r denote $B \cup (\ell.n)$, and let s denote $B \cup (m.n)$. Then $r.m$ and $s.\ell$ join to form a line, call it t. Let t intersect ABC at D. By P6 we see that D is distinct from A, B, C. Then by construction we have $H(A, B; C, D)$.

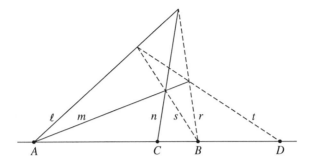

Figure 5.4. Constructing the harmonic conjugate of C w.r.t. A and B.

Now we assume P5, and will prove the uniqueness of the fourth harmonic point. Given A, B, C construct D as above. Suppose D' is another point such that $H(A, B; C, D')$. Then by definition, there is a complete quadrangle $XYZW$ such that $A = XY.ZW$, $B = XZ.YW$, $C \in XW$, $D' \in YZ$. Call $\ell' = AX$, $m' = AZ$, and $n' = CX$. Then we see that the above construction, applied to ℓ', m', n', determines D'.

Thus it is sufficient to show that our construction of D is independent of the choice of ℓ, m, n. We do this in three steps, by showing that if we vary one of ℓ, m, n, the point D remains the same. (Why does this suffice?)

Step 1. If we replace ℓ by a line ℓ', we get the same D.

Let D be defined by ℓ, m, n as above, and label the resulting complete qudrangle $XYZW$. Let ℓ' be another line through A, distinct from m, and label the quadrangle obtained from ℓ', m, n by $X'Y'Z'W$. (Note the point $W = m.n$ belongs to both quadrangles).

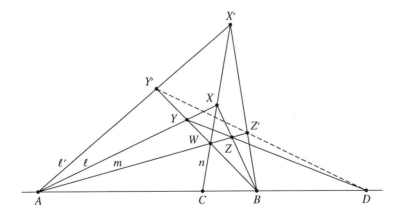

Figure 5.5. Harmonic points, ℓ varies.

We must show that the line $Y'Z'$ passes through D, i.e. that $Y'Z'.AB = D$. Indeed, observe that the two triangles XYZ and $X'Y'Z'$ are perspective from W. Two pairs of corresponding sides meet in A and B respectively: $A = XY.X'Y'$ and $B = XZ.X'Z'$.

Hence by P5, the third pair of corresponding sides, namely YZ and $Y'Z'$, must meet on AB, so that $Y'Z'.AB = D$.

Step 2. If we replace m by m', we get the same D. The proof in this case is identical with that of Step 1, interchanging the roles of ℓ and m.

Step 3. If we replace n by n' we get the same D.

The proof in this case is more difficult, since all four points of the corresponding complete quadrangle change. So let $XYZW$ be the quadrangle formed by ℓ, m, n, which defines D. Let $X'Y'Z'W'$ be the quadrangle formed by ℓ, m, n'. We must show that $Y'Z'$ also meets ABC at D.

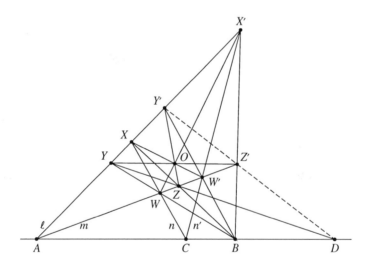

Figure 5.6. Harmonic points, n varies.

Consider the triangles XYW and $W'Z'X'$. Corresponding sides meet in the collinear points A, B, and C, respectively. By D5 the two triangles must be perspective from a point O. In other words, the lines XW', YZ', and WX' meet in a point O.

Similarly, by considering the ordered triangles ZWX and $Y'X'W'$, and applying D5 once more, we deduce that the lines ZY', WX', and XW' are concurrent. Since two of these lines are among the three above, and $XW' \neq X'W$, we conclude that their point of intersection is also O.

In other words, the quadrangles $XYZW$ and $W'Z'Y'X'$ are perspective from O, in that order. In particular, the triangles XYZ and $W'Z'Y'$ are perspective from O. Two pairs of corresponding sides meet in A and B, respectively. Hence the third pair of sides, YZ and $Z'Y'$, must meet on the line AB, i.e. $D \in Z'Y'$.   □

5.8. Proposition *Let AB, CD be four harmonic points. Then (assuming P5) also CD, AB are four harmonic points. Combining with an earlier observation (Definition 5.6), we find therefore*

$$H(A,B;C,D) \Leftrightarrow H(B,A;C,D) \Leftrightarrow H(A,B;D,C) \Leftrightarrow H(B,A;D,C)$$
$$\Updownarrow$$
$$H(C,D;A,B) \Leftrightarrow H(D,C;A,B) \Leftrightarrow H(C,D;B,A) \Leftrightarrow H(D,C;B,A).$$

Proof We assume $H(A,B;C,D)$, and let $XYZW$ be a complete quadrangle as in the definition of harmonic quadruple.

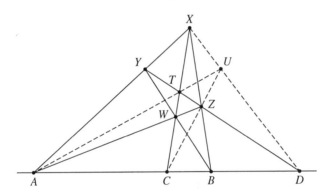

Figure 5.7. Proof of $H(C,D;A,B)$.

Draw DX and CZ, and let them meet in U. Let $XW.YZ = T$. Then $XTUZ$ is a complete quadrangle with C, D as two of its diagonal points: B lies on XZ, so it will be sufficient to prove that TU passes through A. For then we will have $H(C,D;A,B)$.

Consider the two triangles XUZ and YTW. Their corresponding sides meet in D, B, C respectively, which are collinear. Hence by D5, the lines joining corresponding vertices, namely XY, TU, WZ, are concurrent, which is what we wanted to prove.   □

5.9. Example In the projective plane of 13 points, there are 4 points on any line. These 4 points always form a harmonic quadruple, in any order.

To prove this, it will be sufficient to show that P6 holds in this plane. For then there will always be a fourth harmonic point to any 3 points, and it must be *the* fourth point on the line. We will prove this later: The plane of 13 points is the projective plane over the field of 3 elements, which is of characteristic 3. But P6 holds in the projective plane over any field of characteristic $\neq 2$.

5.10. Example In the real Euclidean plane, three collinear points A, B, C may be assigned the following simple invariant under parallel projection:

$$(A, B; C) = \frac{AC}{BC},$$

the ratio of signed segment lengths into which C divides AB. For example, C is the midpoint of line segment $\overline{AB}$ if and only if $(A, B; C) = -1$ (Exercise 1.14). Now the image under central projection of a line segment and its midpoint will not necessarily be another line segment and its midpoint. Thus, the simple ratio above will fail to define a projective invariant, but it turns out that forming the ratio $(A, B; C)/(A, B; D)$ of *four* collinear points A, B, C, D in the Euclidean plane gives a quantity that remains fixed under central projection (Exercise 5.7).

Hence, four collinear points A, B, C, D are assigned a *cross ratio* defined by

$$R_x(A, B; C, D) = \frac{AC}{BC} \bigg/ \frac{AD}{BD}.$$

Then A, B, C and D form a harmonic quadruple if and only if $R_x(A, B; C, D) = -1$ (Exercise 5.2).

5.11. Example $\angle \ell n$ denotes the signed acute angle between ℓ and n. The cross ratio of four concurrent lines ℓ, m, n, o in the Euclidean plane is best dualized as follows:

$$R_x(\ell, m; n, o) = \frac{\sin \angle \ell n}{\sin \angle \ell o} \cdot \frac{\sin \angle mo}{\sin \angle mn}.$$

In this way, the cross ratio of four lines equals the cross ratio of the four points of intersection with any transversal line (Exercise 5.7).

5.3   Perspectivities and Projectivities

5.12. Definition Let ℓ and ℓ' be lines in a projective plane, and let O be a point not on either ℓ or ℓ'. A *perspectivity* of ℓ onto ℓ' with center O is a one-to-one correspondence of the points of one line ℓ with the points of another line ℓ', which is obtained in the following way. For each point $A \in \ell$, let

$$A \mapsto A' = OA.\ell'.$$

In symbols, we write $\ell \stackrel{O}{\barwedge} \ell'$, and we say "$\ell$ is mapped into ℓ' by a perspectivity with center at O," or $ABC \ldots \stackrel{O}{\barwedge} A'B'C' \ldots$, which says "the points A, B, C (of the line ℓ)

are mapped via a perspectivity with center O into the points A', B', C', respectively (of the line ℓ')."

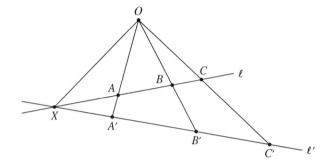

Figure 5.8. A perspectivity between ℓ and ℓ'.

A perspectivity between lines ℓ and ℓ' of course sends $\ell.\ell'$ into itself.

One can see easily that a composition of two or more perspectivities need not be a perspectivity. For example, in Figure 5.9 we have

$$\ell \; \overset{O}{\underset{\wedge}{=}} \; \ell' \; \overset{O'}{\underset{\wedge}{=}} \; \ell'' \qquad \text{and} \qquad ABCY \; \overset{O}{\underset{\wedge}{=}} \; A'B'C'Y' \; \overset{O'}{\underset{\wedge}{=}} \; A''B''C''Y''.$$

Now if the composed map from ℓ to ℓ'' were a perspectivity, it would have to send $\ell.\ell'' = Y$ to itself. However, Y goes into Y''. Therefore we make the following definition.

5.13. Definition A *projectivity* is a mapping of one line ℓ into another ℓ' (which may be equal to ℓ), which can be expressed as a composition of a finite number of perspectivities:

$$\ell \; \overset{O_1}{\underset{\wedge}{=}} \; \ell_1 \; \overset{O_2}{\underset{\wedge}{=}} \; \dots \; \overset{O_{n-1}}{\underset{\wedge}{=}} \; \ell_n \; \overset{O_n}{\underset{\wedge}{=}} \; \ell'$$

We abbreviate this to $\ell \; \overline{\wedge} \; \ell'$, and write $A_1 A_2 \dots A_n \; \overline{\wedge} \; A'_1 A'_2 \dots A'_n$, if the projectivity takes points $A_1, A_2, \dots, A_n$ into $A'_1, A'_2, \dots, A'_n$ respectively.

A projectivity is of course one-to-one and onto.

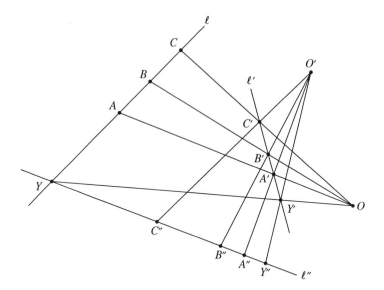

Figure 5.9. A projectivity that is not a perspectivity.

5.14. Proposition *Let ℓ be a line. Then the set of projectivities of ℓ into itself forms a group, which we will call* PJ(ℓ).

> **Proof** The composition of two projectivities is a projectivity, because the result of performing one chain of perspectivities ending in ℓ followed by another starting at ℓ is still a chain of perspectivities. The identity map of ℓ into itself is a projectivity (in fact a perspectivity), and acts as the identity element in PJ(ℓ). The inverse of a projectivity is a projectivity, since we need only reverse the chain of perspectivities. □

Naturally, we would like to study this group. In particular, we would like to know how many times transitive it is. It is clearly 2-transitive, because there exists a group element, viz., a projectivity, sending two arbitrary points, A and B, into two arbitrary points, A' and B' (Exercise 5.4). We will see in the next proposition that it is 3-transitive, and in the succeeding proposition that it cannot be 4-transitive.

5.15. Proposition *In a projective plane π, let A, B, C and A', B', C' be two triples of points on a line ℓ. Then there is a projectivity of ℓ into itself which sends A, B, C into A', B', C'.*

> **Proof** If all six points lie on a line ℓ, we can start arguing as follows. Let ℓ' be a line different from ℓ, which does not pass through A or A'. Let O be any

point not on ℓ, ℓ', and project A', B', C' from ℓ to ℓ', giving A'', B'', C'', so we have

$$A'B'C' \overset{O}{\underset{\wedge}{=}} A''B''C'',$$

and $A \notin \ell'$, $A'' \notin \ell$. Now it is sufficient to construct a projectivity from ℓ to ℓ', taking ABC into $A''B''C''$ (why?). Drop double primes, and forget the original points $A', B', C' \in \ell$. What remains is to do the following problem:

Let ℓ, ℓ' be two distinct lines, let A, B, C be three distinct points on ℓ, and let A', B', C' be three distinct points on ℓ'; assume furthermore that $A \notin \ell'$, and $A' \notin \ell$. Construct a projectivity from ℓ to ℓ' which carries A, B, C into A', B', C', respectively.

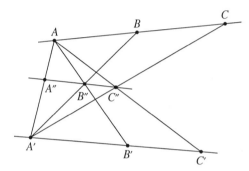

Figure 5.10. A projectivity sending ABC into $A'B'C'$.

Consider lines AA', AB', AC', $A'B$, $A'C$, and let $B'' = AB'.A'B$ and $C'' = AC'.A'C$. Let ℓ'' denote $B''C''$ and meet AA' at A''. Then $\ell \overset{A'}{\underset{\wedge}{=}} \ell'' \overset{A}{\underset{\wedge}{=}} \ell'$ sends

$$ABC \overset{A'}{\underset{\wedge}{=}} A''B''C'' \overset{A}{\underset{\wedge}{=}} A'B'C'.$$

Thus we have found the required projectivity as a composition of two perspectivities.

The case where one or both of A and A' is the point $\ell.\ell'$ is disposed of by relabelling points.   □

5.16. Remark In the presence of Axiom P7, introduced in the next chapter, the line ℓ'' is called the *cross axis* of the projectivity $ABC \overline{\wedge} A'B'C'$. In fact, it does not depend on which three points A, B, C are chosen in the domain of the projectivity (Exercise 6.11).

5.17. Lemma *Suppose $ABCD \overset{O}{\underset{\wedge}{=}} A'BC'D'$ is a perspectivity between distinct lines (with point of intersection B) in a projective plane with P5 and P6. If* $\mathrm{H}(A, B; C, D)$, *then* $\mathrm{H}(A', B; C', D')$.

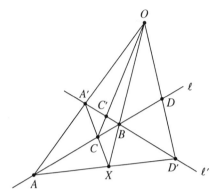

Figure 5.11. Harmonic points A,B,C,D under perspectivity.

Proof Let $X = OB.AD'$. Make note of the lines OAA', OCC' and ODD'. Then $A'OXD'$ is a complete quadrangle with two diagonal points $B = OX.A'D'$ and $A = OA'.XD'$. Since the remaining lines in $A'OXD'$ are OD' and $A'X$, and $OD'.AB = D$, it follows from $H(A,B;D,C)$ and unicity (Proposition 5.7) that $A'X.AB = C$.

Consider now the complete quadrangle $ACXO$. Two of its diagonal points are $A' = CX.OA$ and $B = OX.AC$. The line $A'B$ intersects two other points of the complete quadrangle: $OC.A'B = C'$ and $AX.A'B = D'$. Hence,

$$H(A',B;C',D').$$

□

5.18. Proposition *Assuming P5 and P6, a projectivity sends harmonic points into harmonic points.*

Proof Since a projectivity is a finite composition of perspectivities, it suffices to show that a perspectivity between distinct lines sends harmonic points into harmonic points.

Given $ABCD \overset{O}{\underset{\wedge}{=}} A'B'C'D'$ and $H(A,B;C,D)$, we make use of the line AB' if $B \neq B'$. If $B = B'$, we are done by the lemma. Now

$$ABCD \overset{O}{\underset{\wedge}{=}} AB'C''D'' \overset{O}{\underset{\wedge}{=}} A'B'C'D'$$

where $C'' = OC.AB'$ and $D'' = OD.AB'$. Each of the two perspectivities satisfies the hypothesis of the lemma (that one point of the harmonic quadruple is fixed); hence $H(A,B';C'',D'')$ and $H(A',B';C',D')$ by the lemma.

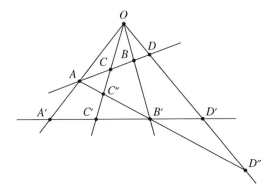

Figure 5.12. Harmonic points A, B, C, D under perspectivity.

$\square$

So we see that the group $PJ(\ell)$ is three times transitive, but is not four times transitive, because it must take quadruples of harmonic points into quadruples of harmonic points.

Exercises

5.1   Find the diagonal points of the complete quadrangle
        (*a*) on the four points $(\pm 1, \pm 1, 1)$,
        (*b*) on the four points $(1, 0, 0)$, $(0, 1, 0)$, $(0, 0, 1)$, and $(1, 1, 1)$.

5.2*   Let π be the real projective plane, and let $A = (a, 0, 1)$, $B = (b, 0, 1)$, $C = (c, 0, 1)$, $D = (d, 0, 1)$, $a, b, c, d \in \mathbf{R}$, be four points on the "*x*-axis". Prove that A, B, C, D are four harmonic points if and only if

$$\mathrm{R}_x(A, B; C, D) = \frac{a-c}{a-d} \cdot \frac{b-d}{b-c} = -1.$$

You may use methods of Euclidean geometry in the affine plane $x_3 \neq 0$.

5.3   If P and Q are points in the real projective plane represented by the vectors v and w, respectively, in $\mathbf{R}^3$, find an expression for the harmonic conjugate with respect to P and Q of the point R represented by $\alpha v + \beta w$.

5.4   Give a simple demonstration that $PJ(\ell)$ is 2-transitive on the set of points constituting ℓ, i.e., given A, B and A', B' on ℓ, find a projectivity $\ell \overline{\wedge} \ell$ taking $A \mapsto A'$, $B \mapsto B'$.

5.5   The tangent lines of the hyperbola $y = 1/x$ provide a means of associating any point on the *x*-axis with a point on the *y*-axis in $\mathbf{R}^2$.
        (*a*) Show that this is the mapping

$$(x, 0) \mapsto \begin{cases} (0, 4/x) & \text{if } x > 0 \text{ or } x < 0; \\ \infty & \text{if } x = 0; \\ (0, 0) & \text{if } x = \infty. \end{cases}$$

(*b*) Show that the projectivity from the *x*-axis into the *y*-axis given by

$$(y=0) \ \overset{(1,1)}{\underset{\wedge}{=}} \ \ell_\infty \ \overset{(1,0)}{\underset{\wedge}{=}} \ (y=x) \ \overset{V}{\underset{\wedge}{=}} \ (y=4x) \ \overset{H}{\underset{\wedge}{=}} \ (x=0),$$

where

$$V = \text{ideal point of all vertical lines}$$
$$H = \text{ideal point of slope 0 lines,}$$

is also the mapping

$$(x,0) \mapsto \begin{cases} (0,4/x) & \text{if } x \neq 0; \\ \infty & \text{if } x = 0; \\ 0 & \text{if } x = \infty. \end{cases}$$

Conclude that the tangent lines of the hyperbola determine a projectivity between the asymptotes.

(*c*) Show that the cross axis of the projectivity in (*a*) and (*b*) is ℓ_∞. (If $ABC \ \overset{}{\underset{\wedge}{}} A'B'C'$, recall that the cross axis is the line $(AB'.BA') \cup (AC'.CA')$.)

5.6   Pick a line ℓ in the seven point plane and compute $PJ(\ell)$. You should arrive at (either a trivial or a proper) subgroup of S_3.

5.7*   Refer to Figure 5.13.

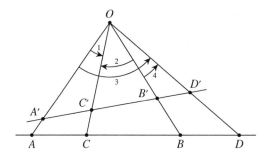

Figure 5.13. Cross ratio is invariant under central projection.

(*a*) Apply the law of sines in the Euclidean plane to show that

$$R_x(A,B;C,D) = \frac{\sin\angle 1}{\sin\angle 2} \Big/ \frac{\sin\angle 3}{\sin\angle 4}.$$

(*b*) Show that

$$R_x(A,B;C,D) = R_x(OA,OB;OC,OD) = R_x(A',B';C',D').$$

(Cf. Example 5.11.)

(*c*) Deduce that projectivities preserve cross ratio in the real projective plane.

5.8   In the Euclidean plane, define the cross ratio of four points on a circle as the cross ratio of the four lines determined by these and concurrent in a fifth point on the circle.

(*a*) Which well-known theorem of Euclidean geometry assures us that this definition is independent of the choice of fifth point?

(*b*) Apply Exercise 5.7 to obtain a good definition of the cross ratio of four points on any conic section.

(You may use this exercise to establish Pascal's theorem in Exercise 6.17.)

5.9   Let ℓ, ℓ' be two distinct lines in a projective plane π. Let $X = \ell.\ell'$. Let A, B be two distinct points on ℓ, different from X. Let C, D be two distinct points on ℓ', different from X. Construct a projectivity $\phi: \ell \to \ell'$ which sends A, X, B into X, C, D, respectively.

5.10   Establish the following: given two harmonic quadruples A, B, C, D and A', B', C', D', there exists a projectivity $ABCD \, \overline{\wedge} \, A'B'C'D'$. Identify the propositions used in your proof.

5.11*   In the ordinary Euclidean plane (considered as being contained in the real projective plane), let C be a circle with center O, let P be a point outside C, and let t_1 and t_2 be the tangents from P to C, meeting C at A_1 and A_2. Draw A_1A_2 to meet OP at B, and let OP meet C at X and Y.

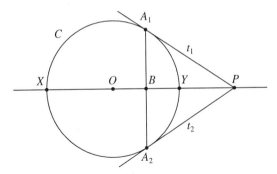

Figure 5.14. Construction of harmonic points using a circle.

(*a*) Prove (by any method) that X, Y, B, P are four harmonic points.

(*b*) What is the harmonic conjugate of the midpoint of a line segment with respect to its endpoints?

(*c*) Show that XY is the *harmonic mean* of XB and XP:

$$\frac{1}{XB} + \frac{1}{XP} = \frac{2}{XY}.$$

5.12   Given a complete quadrangle $ABCD$ with diagonal points $P, Q,$ and R, choose any three points of A, B, C, D, say ABC. Show that the harmonic conjugates of $P, Q,$ and R with respect to $AB, AC,$ and BC lie on a line.[1]

---

[1]Poncelet called this line the *trilinear polar* of D. Much of the theory in this chapter is due to him.

5.13   Use Exercise 5.12 to prove the following Euclidean theorem: the medians of a tri-angle are concurrent.

5.14   Refer to Proposition 5.8. What are the permutations in S_4 under which an ordered 4-tuple of harmonic points remains harmonic? Do they form a group?

5.15*   [2] Suppose that there exist points A, B, C, D, and E on a line in the Euclidean plane such that

$$R_x(A,B;C,D) = R_x(A,B;C,E).$$

(*a*) Show that $D = E$.
(*b*) State and prove the dual.

5.16   In the Euclidean plane one can define X to be between A and B if distance $\text{dist}(A,B) = \text{dist}(A,X) + \text{dist}(X,B)$. This corresponds to our intuitive idea about betweenness. Suppose $AXB \barwedge A'X'B'$. Is X' between A' and B'?

5.17   Let π_7 denote the projective plane of 7 points.
(*a*) Apply Corollary 4.21 to show that every complete quadrangle has collin-ear diagonal points.
(*b*) Show that the hypothesis of Axiom P5 is never satisfied in π_7.

5.18   Prove: Any four collinear points can be interchanged in pairs by a projectivity.
Hint: Show that $ABCD \barwedge BADC$ in figure below. Justify.

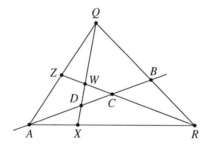

Figure 5.15. Exercise 5.18

---

[2]This exercise extends Proposition 5.7 in the case of the real projective plane. In addition, you should obtain a simpler proof.

Chapter 6

The Fundamental Theorem for Projectivities on a Line

In Proposition 5.15 we saw that, having specified three collinear points A, B, C, and another three collinear points A', B', C', we can find at least one projectivity $ABC \barwedge A'B'C'$. In this chapter, we come to the "fundamental theorem", which states that there is only one projectivity $ABC \barwedge A'B'C'$. Hence the image of a fourth point X on line ABC may be constructed by drawing the line XA' and a line through A and the point of intersection of XA' with the cross axis (cf. Proposition 5.15).

6.1   The Fundamental Theorem: Axiom P7

It turns out that the fundamental theorem cannot be proven from P1–P6, so we introduce it as an additional Axiom P7, on the grounds that it is true in the real projective plane (Theorem 6.6). Then we examine the key role P7 plays in projective geometry: we prove both Pappus' theorem and Desargues' theorem in the presence of P7.

> **P7. Fundamental theorem for projectivities on a line** Let ℓ be a line in a projective plane. Let A, B, C and A', B', C' be two triples of three distinct points on ℓ. Then there is at most one projectivity of ℓ into ℓ such that $ABC \barwedge A'B'C'$.

Since two lines are in one-to-one correspondence by some perspectivity, P7 holds on every line if it holds on a single line. As a simple consequence of P7, $ABC \barwedge ABC$ must be the identity. Then a projectivity not equal to the identity has at most 2 fixed points. Next we consider how P7 may be reformulated in several ways.

61

6.1. Proposition *P7 is equivalent to:*

> **P7′.** Let ℓ and ℓ' be two distinct lines in a projective plane. Let $A, B, C \in \ell$ and $A', B', C' \in \ell'$. Then there is one and only one projectivity of ℓ onto ℓ' such that $ABC \, \overline{\wedge} \, A'B'C'$.
>
> **P7″.** Let ℓ and ℓ' be distinct lines in a projective plane and $X = \ell.\ell'$. Then every projectivity $\ell \, \overline{\wedge} \, \ell'$ sending X to itself is a perspectivity.

Proof P7 $\implies$ P7′. Suppose two distinct projectivities $\phi\colon ABC \, \overline{\wedge} \, A'B'C'$ and $\psi\colon ABC \, \overline{\wedge} \, A'B'C'$ exist mapping ℓ onto ℓ'. Then there is $X \in \ell$ such that $\phi(X) \neq \psi(X)$. Let O be a point not on either ℓ or ℓ'. Denote the perspectivity with center O between ℓ' and ℓ by $\tau\colon A'B'C' \, \overset{O}{\overline{\wedge}} \, A''B''C''$ where $A'' = OA'.\ell$, etc. Then $\tau\phi$ and $\tau\psi$ are two projectivities $ABC \, \overline{\wedge} \, A''B''C''$ of ℓ onto itself, but $\tau(\phi(X)) \neq \tau(\psi(X))$ since τ is a one-to-one correspondence. This contradicts P7.

  P7′ $\implies$ P7″. Let A, B be two points on ℓ different from $X = \ell.\ell'$, and A', B' their images under a projectivity ϕ, which sends X to itself. Let $O = AA'.BB'$. Then ϕ must equal the perspectivity $ABX \, \overset{O}{\overline{\wedge}} \, A'B'X$ by P7′.

  P7″ $\implies$ P7. Suppose we have two projectivities on a line ℓ'' with the same effect on a triple of points: $\psi_1, \psi_2\colon PQR \, \overline{\wedge} \, P'Q'R'$. Given distinct lines ℓ and ℓ' we construct projectivities $\phi_1\colon PQR \, \overline{\wedge} \, XAB$ and $\phi_2\colon P'Q'R' \, \overline{\wedge} \, XA'B'$ where $X = \ell.\ell'$, $A, B \in \ell$ and $A', B' \in \ell'$. Then both $\phi_2\psi_1\phi_1^{-1}$ and $\phi_2\psi_2\phi_1^{-1}$ send $XAB \, \overline{\wedge} \, XA'B'$. By P7″, $\phi_2\psi_1\phi_1^{-1} = \phi_2\psi_2\phi_1^{-1}$, since both functions are perspectivities with center $AA'.BB'$. Multiplying by ϕ_2^{-1} from the left and ϕ_1 from the right results in $\psi_1 = \psi_2$.[1] $\square$

The principle of duality would naturally provide us with perspectivities between pencils of lines via a line as axis. We extend the principle of duality with the next proposition.

6.2. Proposition *P7 implies its dual statement:*

> **D7.** *Let P be a point in a projective plane. Let a, b, c and a′, b′, c′ be two triples of lines through P. Then there exists one and only one projectivity abc $\overline{\wedge}$ a′b′c′.*

---

[1] With the same sort of argument one establishes the following principle: if in a projective plane π there exist collinear points A, B, C and A', B', C' such that $\psi_1, \psi_2\colon ABC \, \overline{\wedge} \, A'B'C' \implies \psi_1 = \psi_2$, then π satisfies P7 (Exercise 6.3). In other words, the fundamental theorem is equivalent to any of its special cases.

Proof An *elementary correspondence* τ gives a one-to-one correspondence from a pencil of lines through Q, to the set of points, or *range* of points, on a line ℓ not passing through Q by associating a line m through P with the point $\ell.m$. (τ^{-1} is the same elementary correspondence, but the inverse mapping, from a range of points to a pencil of lines. A perspectivity is then the application of two elementary correspondences.)

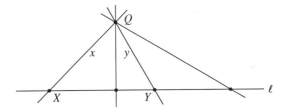

Figure 6.1. An elementary correspondence, $xy \stackrel{}{\overline{\wedge}} XY$.

Now let P be the point in the hypothesis and ℓ a line with $P \notin \ell$. Consider the elementary correspondence τ from the pencil of lines at P to the range of points on ℓ. Let $\tau(a) = A$, $\tau(b) = B, \dots , \tau(c') = C'$. If $\psi_1, \psi_2 : abc \stackrel{}{\overline{\wedge}} a'b'c'$ are two projectivities, then $\tau\psi_1\tau^{-1}, \tau\psi_2\tau^{-1} : ABC \stackrel{}{\overline{\wedge}} A'B'C'$ are two projectivities between ranges of points. (Why are $\tau\psi_i\tau^{-1}$ projectivities? Exercise 6.8). We have $\tau\psi_1\tau^{-1} = \tau\psi_2\tau^{-1}$ by P7, whence $\psi_1 = \psi_2$ as mappings, which establishes D7. $\qquad\qquad\square$

The next theorem says that *the fundamental theorem implies Desargues' theorem* (or briefly P7 $\Longrightarrow$ P5).

6.3. Theorem *If P7 holds in a projective plane π, then P5 holds in π.*

Proof Given lines $OAA', OBB', OCC', ABP, ACQ, BCR, A'B'P, A'C'Q, B'C'R$, we have to show that PQR is a line. We will prove that P is on QR.

Let $S = CP.QR$, $T = A'B'.C'S$, $X = AB.OC$, $Y = OC.QR$, $Z = OC'.A'B'$. We consider first the projectivity

$$ABXP \stackrel{O}{\overline{\wedge}} A'B'ZP.$$

Second, consider the projectivity

$$ABXP \stackrel{C}{\overline{\wedge}} QRYS \stackrel{C'}{\overline{\wedge}} A'B'ZT.$$

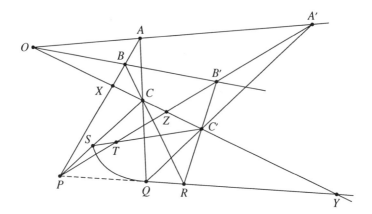

Figure 6.2. Proof of Desargues' theorem: $P = T = S$.

By the fundamental theorem, the two projectivities must be identical, hence $T = P$. This implies that $P \in C'S$. Since $S \in CP$, we get $S \in C'P.CP$. So $S = P$, which implies that $P \in QR$. $\square$

6.2   Geometry of Complex Numbers

In this section, we wish to check P7 in the real projective plane. In order to do this, we introduce Möbius transformations of the complex numbers and prove a lemma, all of which will be useful also in later chapters. An alternative proof that $\mathbf{P}^2(\mathbf{R})$ satisfies P7 is sketched in Exercise 6.5.

The real Cartesian plane $\mathbf{R}^2$ may be identified with the complex numbers $\mathbf{C}$ via the mapping $(a,b) \mapsto a + bi$, thereby giving $\mathbf{R}^2$ multiplication and division operations

$$(a,b)(c,d) = (ac - bd, bc + ad)$$

and

$$(a,b)/(c,d) = \left(\frac{ac + bd}{c^2 + d^2}, \frac{bc - ad}{c^2 + d^2} \right),$$

so long as $c^2 + d^2 \neq 0$. These formulas are usually remembered by using the complex number notation and the relation $i^2 = -1$.

6.4. Definition Let $\mathbf{C}_\infty$ denote the *extended complex numbers* $\mathbf{C} \cup \{\infty\}$. Define for all $a, b, c, d \in \mathbf{C}$ such that $ad - bc \neq 0$ a mapping $f: \mathbf{C}_\infty \to \mathbf{C}_\infty$ by

$$x \mapsto \frac{ax + b}{cx + d}, \qquad x \mapsto \begin{cases} \infty & \text{when } x = \infty, c = 0; \\ a/c & \text{when } x = \infty, c \neq 0; \\ \infty & \text{when } x = -d/c, c \neq 0. \end{cases}$$

f is called a *Möbius transformation*, or *fractional linear transformation*.

6.5. Lemma *The set of Möbius transformations form a group under composition; in particular, each Möbius transformation is a one-to-one correspondence of $\mathbf{C}_\infty$ with itself. Furthermore, a Möbius transformation is determined by its value on three elements in $\mathbf{C}_\infty$.*

Proof Given a Möbius transformation denoted by $f(x) = \frac{ax+b}{cx+d}$, and another $g(x) = \frac{a'x+b'}{c'x+d'}$, their composite is

$$f \circ g(x) = \frac{(aa' + bc')x + (ab' + bd')}{(ca' + dc')x + (cb' + dd')}$$

as can be obtained by direct computation. Note that the coefficients in the composite correspond from left to right, up to down, with the coefficients in the matrix product

$$\begin{pmatrix} a & b \\ c & d \end{pmatrix} \begin{pmatrix} a' & b' \\ c' & d' \end{pmatrix} = \begin{pmatrix} aa' + bc' & ab' + bd' \\ ca' + dc' & cb' + dd' \end{pmatrix},$$

while the condition $ad - bc \neq 0$ is equivalent to $\det \left(\begin{smallmatrix} a & b \\ c & d \end{smallmatrix} \right) \neq 0$. Since determinant of square matrices A, B satisfies $\det AB = \det A \det B$, it follows that $f \circ g$ is a Möbius transformation.

The identity $f(x) = x$ is obtained when $a = d = 1, b = c = 0$. Moreover, a Möbius transformation $\frac{ax+b}{cx+d}$ has inverse function $\frac{dx-b}{-cx+a}$ corresponding to the adjugate matrix: whence it is a one-to-one correspondence of $\mathbf{C}_\infty$ with itself.

To prove the last statement in the lemma it suffices to check that there is one and only one Möbius transformation $f_{\alpha,\beta,\gamma}$ taking $0, 1, \infty$ to an arbitrary triple α, β, γ in $\mathbf{C}_\infty$. For then

$$f_{\alpha',\beta',\gamma'} \circ f_{\alpha,\beta,\gamma}^{-1}$$

is clearly the unique Möbius transformation sending an arbitrary triple α, β, γ to α', β', γ' in $\mathbf{C}_\infty$.

First suppose $\alpha, \beta, \gamma \in \mathbf{C}$ and $f(x) = \frac{ax+b}{cx+d}$. Then

$$f(0) = \frac{b}{d} = \alpha, \quad f(1) = \frac{a+b}{c+d} = \beta, \quad f(\infty) = \frac{a}{c} = \gamma.$$

Clearly, one parameter of a Möbius transformation is freely chosen, so take $d = 1$. This forces $b = \alpha$, $a = c\gamma$, and $\beta = \frac{a+b}{c+1}$, so $\beta = \frac{c\gamma+\alpha}{c+1}$, forcing $c = \frac{\alpha-\beta}{\beta-\gamma}$ and $a = \gamma\frac{\alpha-\beta}{\beta-\gamma}$. Then

$$ad - bc = \frac{(\alpha - \beta)(\gamma - \alpha)}{\beta - \gamma} \neq 0;$$

after multiplying through by the factor $\beta - \gamma$, we get

$$f(x) = \frac{\gamma(\alpha - \beta)x + \alpha(\beta - \gamma)}{(\alpha - \beta)x + (\beta - \gamma)}$$

to be the Möbius transformation determined by the three equations. If one of $\alpha, \beta, \gamma = \infty$, a similar computation determines a unique Möbius transformation (Exercise 6.12).   □

6.6. Theorem *The fundamental theorem (Axiom P7) holds in the real projective plane.*

Proof We first show that a perspectivity $\ell \overset{O}{\barwedge} \ell'$ is given by a Möbius transformation if the ideal points on ℓ and ℓ' are both identified with ∞. Let ℓ be parametrized by $(a + ib)t + c + id = ut + v$, and ℓ' by $(a' + ib')s + c' + id' = u's + v'$, and let O be given by coordinates (p, q).

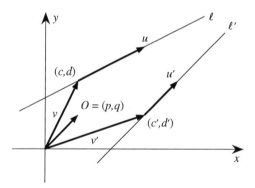

Figure 6.3. $u = a + ib$ and $u' = a' + ib'$

We claim that $\ell \overset{O}{\barwedge} \ell'$ is the Möbius transformation $M_4 M_3^{-1} M_2 M_1^{-1}(t)$ of $\mathbf{C}_\infty$, restricted to $\ell \cup \infty$, where M_1, M_2, M_3 and M_4 are given by

$$M_1(x) = ux + v, \qquad M_2(x) = \frac{ax + c - p}{bx + d - q}$$

$$M_3(x) = \frac{a'x + c' - p}{b'x + d' - q}, \qquad M_4(x) = u'x + v'.$$

Since $M_4 M_3^{-1} M_2 M_1^{-1}(ut + v) = u' M_3^{-1} M_2(t) + v'$, we see that $\ell \cup \infty$ is mapped into $\ell' \cup \infty$. Then it suffices to show that $ut + v$, $p + iq$, and $u' M_3^{-1} M_2(t) + v'$ are collinear. Let $M = M_3^{-1} M_2$. We must show that the slope of the line through

points $ut + v$ and $p + iq$ equals the slope of the line through points $u'M(t) + v'$
and $p + iq$:

$$\frac{bt + d - q}{at + c - p} = \frac{b'M(t) + d' - q}{a'M(t) + c' - p}.$$

Now, the left hand side is equal to $1/M_2(t)$, the right hand side equal to

$$\frac{1}{M_3(M(t))} = \frac{1}{M_3(M_3^{-1}M_2(t))} = \frac{1}{M_2(t)}.$$

This proves the identity. Thus a perspectivity $\ell \overset{O}{\barwedge} \ell'$ with center O in $\mathbf{R}^2$ is a
Möbius transformation of the line ℓ in $\mathbf{C}_\infty$ onto ℓ' in $\mathbf{C}_\infty$. It is left as an easy
exercise to check the same if O is an ideal point (giving parallel projection of
ℓ onto ℓ'; this implies $M_3^{-1}M_2(t) = kt$ for some stretching constant k). It follows that any projectivity $\ell \barwedge \ell'$ is similarly a complex Möbius transformation
restricted to $\ell \cup \{\infty\}$. By Lemma 6.5, a Möbius transformation is determined
by its effect on three points. Hence, any projectivity is uniquely determined
by its values on three points, which is the fundamental theorem.   □

6.3   Pappus' Theorem

We now come to one of the oldest of projective theorems, which states that if six
vertices of a hexagon lie alternately on two lines, then the three pairs of opposite
sides meet in collinear points. The theorem was discovered by Pappus of Alexandria, living in the fourth century A.D., and demonstrated with laborious Euclidean
methods.

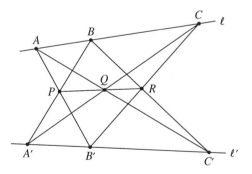

Figure 6.4. Pappus' theorem (alternative formulation): if hexagon $AB'CA'BC'$ is inscribed
on two lines, then the pairs of opposite sides meet in three collinear points.

6.7. Pappus' theorem *Let ℓ and ℓ' be distinct lines in a projective plane with P7.
Let A, B, C be three distinct points on ℓ, different from $Y = \ell.\ell'$. Let A', B', C' be*

three distinct points on ℓ', different from Y. Define $P = AB'.A'B$, $Q = AC'.A'C$, and $R = BC'.B'C$. Then P, Q, and R are collinear.

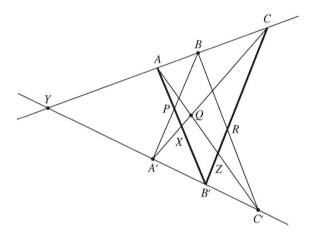

Figure 6.5. Proof of Pappus' theorem.

Proof Let $X = AB'.A'C$ and $Z = AC'.B'C$. Consider the projectivity

$$AXB'P \underset{\wedge}{\overset{A'}{=}} ACYB \underset{\wedge}{\overset{C'}{=}} ZCB'R.$$

Since B' is fixed, and is the point of intersection of the first and last line, P7″ states that this projectivity is a perspectivity. Its center may be computed as the point of intersection $AZ.XC = AC'.A'C = Q$. Hence

$$AXB'P \underset{\wedge}{\overset{Q}{=}} ZCB'R.$$

The last points in perspective tell us that P, Q, and R lie on a line.   □

The last theorem states in brief that P7 implies Pappus. The converse, Pappus implies P7, is in fact true. The interested reader will be led through the proof in Exercises 6.13, 6.14–6.16 below. Thus, Pappus' theorem is equivalent to the fundamental theorem (P7), and might have been used as an axiom in place of P7. For this reason, projective planes that satisfy P7 are referred to as *Pappian planes*.

Exercises

6.1   If A, B, C and A', B', C' are two triples on the same line, construct the image X' of a fourth collinear point X under the projectivity $ABC \underset{\wedge}{\overline{\wedge}} A'B'C'$ (assuming the Fundamental theorem).

6.2   Let π be a finite Pappian plane having $p^2 + p + 1$ points in all. If ℓ is a line in π, then what is the order of the group $PJ(\ell)$?

6.3   Suppose that the following statement is true in π: there exist collinear triples A, B, C and A', B', C' such that if ψ_1, ψ_2 both send $ABC \overline{\wedge} A'B'C'$, then $\psi_1 = \psi_2$ as mappings. Show that Axiom P7 holds in π: i.e. if P, Q, R and P', Q', R' are any two triples of points on a line ℓ, then there is a unique projectivity $PQR \overline{\wedge} P'Q'R'$.

6.4   If A, B, C and A', B', C' are triples of points on a line ℓ in a projective plane π, show that if there are two different projectivities $ABC \overline{\wedge} A'B'C'$, then there exist two different projectivities $ABC \overline{\wedge} A''B''C''$ between distinct lines of π.

6.5*   (Together with Exercises 5.7 and 5.15, this exercise will provide an alternative proof that P7 holds in the real projective plane.) Suppose ψ_1 and ψ_2 are projectivities on a real projective line both sending $ABC \overline{\wedge} A'B'C'$. Choose a fourth point X, different from A, B, or C and show that $\psi_1(X) = \psi_2(X)$ by using cross ratios.

6.6   For each of the following projective planes, state which of the axioms P5, P6, P7 holds in it, and explain why each axiom does or does not hold. (Please refer to results proved earlier, and give *brief* outlines of their proofs.)

> (*a*) The projective plane of seven points.
> (*b*) The real projective plane.
> (*c*) The Moulton plane.

6.7   Let π be a projective plane satisfying P5, P6, and P7, and let ℓ be a line in π.
> (*a*) Prove that if ϕ is a projectivity of ℓ onto ℓ which interchanges two distinct points A, B of ℓ (i.e. $\phi(A) = B$ and $\phi(B) = A$), then ϕ^2 is the identity.[2]
> (*b*) Conclude that $A'B'C'D' \overline{\wedge} B'A'D'C'$ for any 4 points A', B', C', D' on a line.

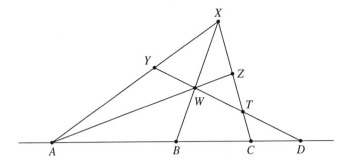

Figure 6.6. Existence of involution $ABCD \overline{\wedge} BADC$.

Hint: Let C be another point of ℓ and let $\phi(C) = D$. Construct a projectivity $\phi' : \ell \to \ell$ which interchanges A and B, and interchanges C and D, using Figure 6.6; justify. Then apply the fundamental theorem.

---

[2] A projectivity $\phi : \ell \overline{\wedge} \ell$ such that $\phi^2 = \mathrm{id}_\ell$ is called an *involution*. You are asked to prove that a projectivity interchanging a pair of points is an involution.

6.8   Suppose τ is the elementary correspondence between the pencil $[P]$ of lines through P and range of points on ℓ. Show that if ϕ is a projectivity between the pencil of lines,

$$[P] \overset{\ell_1}{\underset{\wedge}{=}} [P_1] \overset{\ell_2}{\underset{\wedge}{=}} \cdots \overset{\ell_n}{\underset{\wedge}{=}} [P],$$

then $\tau\phi\tau^{-1}$ is a projectivity between the range of points

$$\ell \overset{P}{\underset{\wedge}{=}} \ell_1 \overset{P_1}{\underset{\wedge}{=}} \ell_2 \overset{P_2}{\underset{\wedge}{=}} \cdots \overset{P_{n-1}}{\underset{\wedge}{=}} \ell_n \overset{P}{\underset{\wedge}{=}} \ell.$$

6.9   Let T be a complex Möbius transformation sending three points Z, Z_1, Z_2 on a line ℓ in $\mathbf{R}^2$ into three points Z, W_1, W_2, respectively, on a line m. Notice that $\ell.m = Z$.

(a) Is T restricted to ℓ a perspectivity from ℓ to m? If so, where is the center?

(b) How many fixed points does T have?

6.10   Establish directly Pappus' theorem in the real projective plane:

(a) Let $\ell, \ell', A, B, C, A', B', C'$ be as in the statement of Pappus' theorem, and take ℓ to be the line at infinity. Then prove by Euclidean methods that $P = A'B.AB'$, $Q = A'C.AC'$, and $R = B'C.BC'$ are collinear.

(b) You may instead solve the following problem, which also shows Pappus' theorem true in the real projective plane: Let ℓ and ℓ' be lines in the Euclidean plane, such that $A, B, C \in \ell, A', B', C' \in \ell'$ and $P = AB'.A'B, Q = AC'.A'C$ are ideal points. Show by methods of Euclidean geometry (similar triangles, parallel lines, corresponding angles, and parallelograms) that $R = BC'.B'C$ is an ideal point.

6.11   Prove that the cross axis does not depend on which three points A, B, C are chosen in the domain of the projectivity (cf. Remark 5.16).

6.12   Prove the existence and uniqueness of a Möbius transformation sending $0, 1, \infty$ to α, β, γ in each case where one of α, β, γ is ∞.

6.13   **G. Hessenberg, 1905** Prove that Pappus' theorem implies Desargues' theorem.

Hint: Making use of the figure below, define $S = A'C'.AB$ and apply Pappus' theorem to the triad $\left(\begin{smallmatrix} O & C & C' \\ B & S & A \end{smallmatrix}\right)$, find the new points, and apply Pappus' theorem again to the triad $\left(\begin{smallmatrix} O & B & B' \\ C & A' & S \end{smallmatrix}\right)$. Find the new point and apply Pappus' theorem again to conclude that P, Q, R are collinear. (Finally, check that your argument does not depend on appearances in the diagram.)

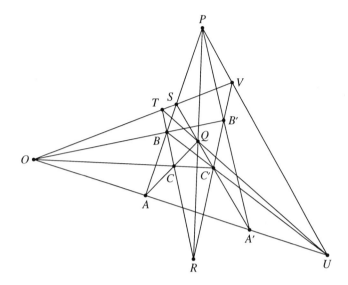

Figure 6.7. Pappus' theorem implies Desargues' theorem.

Exercises 6.14–6.16. Suppose π is a projective plane where Pappus' theorem is true (and hence Desargues' theorem). The next three exercises will guide you through to a proof that π satisfies P7. Thus, Pappus' theorem is equivalent to the fundamental theorem, since Theorem 6.7 tells us that FT $\implies$ Pappus. As pointed out in the text, it will suffice to show that any projectivity between distinct lines, $\tau\colon \ell \to \ell'$, which fixes $X = \ell.\ell'$, i.e. $\tau(X) = X$, is a perspectivity.

6.14 Suppose ℓ_1, ℓ_2, and ℓ_3 are distinct concurrent lines. Suppose the projectivity τ is given by $\ell_1 \overset{P_1}{\underset{\wedge}{=}} \ell_2 \overset{P_2}{\underset{\wedge}{=}} \ell_3$. Let the effect of τ on points be given by

$$A_1X_1 \overset{P_1}{\underset{\wedge}{=}} A_2X_2 \overset{P_2}{\underset{\wedge}{=}} A_3X_3.$$

Use Desargues' theorem to prove that ϕ is the perspectivity $\ell_1 \overset{Q}{\underset{\wedge}{=}} \ell_3$ where $Q = A_1A_3.P_1P_2$.

6.15 Suppose τ is the projectivity given by

$$\ell \overset{P_1}{\underset{\wedge}{=}} \ell_1 \overset{P_2}{\underset{\wedge}{=}} \ell_2 \overset{P_3}{\underset{\wedge}{=}} \ell',$$

where $\ell, \ell_1, \ell_2, \ell'$ are distinct lines, no three being concurrent. Consider the line m joining $\ell.\ell_1$ and $\ell_2.\ell'$. Aside from the case where $P_2 \in m$, τ is equal to the projectivity

$$\ell \overset{P_1}{\underset{\wedge}{=}} \ell_1 \overset{P_2}{\underset{\wedge}{=}} m \overset{P_2}{\underset{\wedge}{=}} \ell_2 \overset{P_3}{\underset{\wedge}{=}} \ell'.$$

(a) Show that points Q and R may be found such that τ is the projectivity $\ell \overset{Q}{\underset{\wedge}{=}} m \overset{R}{\underset{\wedge}{=}} \ell'$.

(*b*) Show that the chain of perspectivities defining τ may also be shortened by one perspectivity in the exceptional case where $P_2 \in m$.

(*c*) Now assume that ℓ, ℓ_1, and ℓ' are concurrent. By inserting a line through $\ell_1.\ell_2$ and avoiding P_3, reduce to the case we started with (four lines, no three concurrent).

(*d*) Conclude that any projectivity between distinct lines may be given as a chain of only two perspectivities.

6.16   By Exercise 6.15, we may find points $P_1, P_2 \in \pi$ and line m, such that τ is the projectivity

$$\ell \; \overset{P_1}{\underset{\wedge}{=}} \; m \; \overset{P_2}{\underset{\wedge}{=}} \; \ell'.$$

If m is concurrent with ℓ and ℓ', we are done by Exercise 6.14. Suppose m is not concurrent with ℓ and ℓ'. Recall that $X = \ell.\ell'$ and $\tau(X) = X$. Let $\ell.m = V$, $\ell'.m = U$, and $XP_1.m = T$ (Figure 6.8).

(*a*) Show that P_1, P_2, and X are collinear.

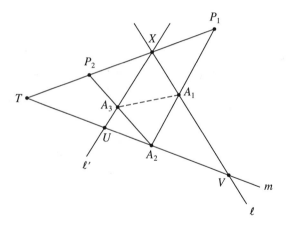

Figure 6.8.

(*b*) Let τ map a point $A_1 \mapsto A_2 \mapsto A_3$ as pictured. Applying Pappus, show that $A_1 A_3$ passes through the point $Q = P_1 U.P_2 V$ for all $A_1 \in \ell$ with image $A_3 \in \ell'$ under τ. Conclude that $\tau \colon \ell \; \overset{Q}{\underset{\wedge}{=}} \; \ell'$, and that π satisfies Axiom P7.

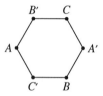

Figure 6.9. The regular hexagon $AB'CA'BC'$.

6.17   Prove Pascal's theorem in $\mathbf{P}^2(\mathbf{R})$: *If hexagon $AB'CA'BC'$ is inscribed in a conic section, then the three pairs of opposite sides meet in three collinear points.*

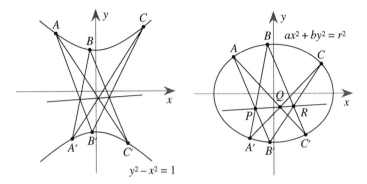

Figure 6.10. Pascal's theorem.

Hint: Make use of the cross ratio of 4 points on a conic section, which you have defined in Exercise 5.8, and uniqueness of the fourth line in fixed cross ratio (the dual of Exercise 5.15). Prove that lines PR and QR are equal by using the proof of Pappus' theorem, but letting equality of cross ratio replace perspectivity.

6.18   Show that complex Möbius transformations map circles and lines into circles and lines.

6.19   **Castillon's problem.** Given a circle Γ in the Euclidean plane and points $A, B, C \notin \Gamma$, when is there a triangle inscribed on Γ whose sides pass through A, B, and C?
      Hint: Transform the circle into a line, using Möbius transformations, and restate the problem.

6.20   Refer to Figure 6.11. Find two lines, label three points A, B, C on one, and three points A', B', C' on the other, such that one application of Pappus' theorem will show that q, r, and p are concurrent. Conclude that Pappus' theorem implies its dual.

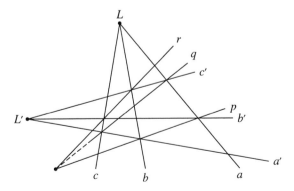

Figure 6.11. The dual of Pappus.

Chapter 7

A Brief Introduction to Division Rings

In this chapter we introduce the notions of ring, division ring and field. We give basic examples of each: polynomial and Laurent series rings, quaternions and the finite fields of integers modulo p. Replacing the field of real numbers $\mathbf{R}$ in $\mathbf{P}^2(\mathbf{R})$ with general division rings will provide us with many examples of projective planes in Chapter 8.

7.1 Division Rings

7.1. Definition A *ring* is a set R together with two binary operations, an "addition" $+$ and a "multiplication" $\cdot$ on R, denoted by $(a,b) \mapsto a+b$ and $(a,b) \mapsto ab$, such that

> **R1.** $(R,+)$ is an abelian group, whose neutral element we denote by 0.
> **R2.** $(R,\cdot)$ is a semigroup[1] possessing a unit element 1.
> **R3.** Multiplication is left and right distributive over addition:

$$a(b+c) = ab+ac, \qquad (a+b)c = ac+bc \qquad (\forall a,b,c \in R).$$

A *division ring* is a ring R such that

> **DR1.** $(R \smallsetminus \{0\}, \cdot, 1)$ is a group.

I.e., every nonzero element has a multiplicative inverse.

---
[1] We have followed tradition and dropped the dot in $a \cdot b$, although we use the dot to denote the binary operation of multiplication.

75

A *field* is a division ring F such that

F1. $ab = ba$ (for all $a, b \in F$)

I.e., multiplication is a commutative operation.

7.2. Example The set of integers $\mathbf{Z}$ under addition and multiplication is clearly a ring. Since multiplication is a commutative operation, we can express this by saying that $\mathbf{Z}$ is a *commutative ring*.

7.3. Example The set of all $n \times n$ matrices ($n > 1$) with real coefficients, $M_n(\mathbf{R})$, is a ring under the usual matrix addition and matrix multiplication. In this example multiplication is noncommutative, and not every nonzero matrix has an inverse. Other examples of rings are obtained by replacing $\mathbf{R}$ in $M_n(\mathbf{R})$ with $\mathbf{Q}, \mathbf{C}$, any field or indeed any ring (Exercise 7.3).

7.4. Example Familiar examples of fields are the rationals $\mathbf{Q}$, the reals $\mathbf{R}$, the complex numbers $\mathbf{C}$ under ordinary addition and multiplication.

A *subfield* is straightforwardly defined as a subset of a field closed under addition and multiplication, containing 0 and 1, and itself a field. For example, $\mathbf{Q}$ is a subfield of $\mathbf{R}$, which in turn is a subfield of $\mathbf{C}$.

7.5. The integers modulo n Let n be an integer greater than 1. We say integers a and b are *congruent modulo n*, writing $a \equiv b \pmod{n}$, if $a - b$ is a multiple of n. Check that this is an equivalence relation on $\mathbf{Z}$ (Exercise 7.4).

Denote the equivalence class of a by $[a]$, i.e.

$$[a] = \{b \in \mathbf{Z} \mid b \equiv a \pmod{n}\},$$

and the set of all equivalence classes by $\mathbf{Z}_n = \{[0], [1], \ldots, [n-1]\}$. Note that $[a]$ is the left coset $a + n\mathbf{Z}$ of the subgroup $n\mathbf{Z}$ of $\mathbf{Z}$ under addition; according to Exercise 4.4, $\mathbf{Z}_n$ is a group with addition given by

$$[a] + [b] = [a + b].$$

By Exercise 4.6 the map $\pi: a \mapsto [a]$ is a surjective group homomorphism. Define a *ring homomorphism* to be a mapping between rings $f: R_1 \to R_2$ satisfying

$$f(x + y) = f(x) + f(y), \qquad f(xy) = f(x)f(y) \qquad (\forall x, y \in R_1).$$

Now you might guess that decreeing π a ring homomorphism would carry over the multiplication to $\mathbf{Z}_n$ making it a ring, and you would be right. Indeed, define a multiplication in $\mathbf{Z}_n$ by

$$[a][b] = [ab].$$

This definition is independent of the representatives a and b of their respective equivalence classes, since

$$(a+kn)(b+rn) = ab + (kb + ra + krn)n.$$

It is a simple exercise to check commutativity of multiplication, and distributivity of multiplication over addition in $\mathbf{Z}_n$. Hence $\mathbf{Z}_n$ is a commutative ring.

Let $n = 6$: observe that $[2][3] = [0]$ in $\mathbf{Z}_6$, so clearly $\mathbf{Z}_6$ is not a field. (In a general ring R, an element a is called a *zero divisor* if $a \neq 0$ and $ab = 0$ for some $b \in R \setminus \{0\}$. A zero divisor cannot have an inverse (Exercise 7.1c).

Now suppose n is a prime p. We claim that $\mathbf{Z}_p$ is a field, i.e., if $[a] \neq [0]$, then $[a]$ is invertible. We see why from this elementary fact about the integers: since a and p are relatively prime, there exist integers b and k such that $ab + pk = 1$ (Exercise 7.6). Then $[a][b] = [1]$, so $[a]$ has an inverse as claimed.

Note that in $\mathbf{Z}_p$, adding $[1]$ together to itself p times gives 0:

$$p[1] = [0].$$

We say that $\mathbf{Z}_p$ is a field of characteristic p, and make the following definition.

7.6. Definition Let F be a division ring. The *characteristic* $\operatorname{char} F$ of F is the smallest integer $p \geq 2$ such that

$$\overbrace{1 + \cdots + 1}^{p \text{ times}} = 0$$

or, if there is no such integer, the characteristic of F is defined to be 0.

7.7. Proposition *The characteristic p of a division ring F is always 0 or a prime number.*

Proof If $p \neq 0$, suppose $p = m \cdot n$ where $m, n > 1$. Then

$$(m1)(n1) = \overbrace{(1 + \cdots + 1)}^{m \text{ times}}\overbrace{(1 + \cdots + 1)}^{n \text{ times}} = p1 = 0.$$

Hence one of $m1$ or $n1$ must be 0, which contradicts the choice of p as the smallest such number (and see Exercise 7.12). Hence, p is prime. □

7.8. Examples of characteristic The fields $\mathbf{Q}$, $\mathbf{R}$, and $\mathbf{C}$ have characteristic 0. The fields $\mathbf{Z}_p$ show that there exists a field of each prime characteristic. In an algebra course, one learns that for each prime power p^n, there exists a field of order p^n and characteristic p. The quaternions described below form a noncommutative division ring of characteristic 0. Later we construct a skew Laurent series ring over $\mathbf{Z}_p(x)$, an example of a noncommutative division ring of characteristic p.

7.2   The Quaternions H

We define a division ring, denoted by **H**, [2] and called the *quaternions*, on the set $\mathbf{R}^4$ of ordered 4-tuples of reals. Addition is just coordinatewise addition familiar from linear algebra:

$$(x_1, x_2, x_3, x_4) + (y_1, y_2, y_3, y_4) = (x_1 + y_1, x_2 + y_2, x_3 + y_3, x_4 + y_4).$$

Multiplication is given by the complicated but interesting formula;

$$\begin{aligned}(x_1, x_2, x_3, x_4) \cdot (y_1, y_2, y_3, y_4) = \\
(x_1 y_1 - x_2 y_2 - x_3 y_3 - x_4 y_4, x_1 y_2 + x_2 y_1 + x_3 y_4 - x_4 y_3, \\
x_1 y_3 + x_3 y_1 + x_4 y_2 - x_2 y_4, x_1 y_4 + x_4 y_1 + x_2 y_3 - x_3 y_2).\end{aligned} \quad (7.1)$$

The formula does lend itself to four quick insights.

(1) $(1, 0, 0, 0)$ is the unit element.

(2) Multiplication is distributive over addition.

(3) We have

$$\begin{aligned}(cx_1, cx_2, cx_3, cx_4) \cdot (y_1, y_2, y_3, y_4) = \\
(x_1, x_2, x_3, x_4) \cdot (cy_1, cy_2, cy_3, cy_4) \qquad (\forall c \in \mathbf{R}).\end{aligned}$$

Using ordinary scalar multiplication of $\mathbf{R}$ in $\mathbf{R}^4$, we write at times $c(x_1, x_2, x_3, x_4)$ in place of (cx_1, cx_2, cx_3, cx_4).

(4) If

$$|(x_1, x_2, x_3, x_4)| = \sqrt{x_1^2 + x_2^2 + x_3^2 + x_4^2}$$

denotes the Euclidean norm on $\mathbf{R}^4$, then

$$\begin{aligned}(x_1, x_2, x_3, x_4) \cdot (x_1, -x_2, -x_3, -x_4) = (|(x_1, x_2, x_3, x_4)|^2, 0, 0, 0) \\
= (x_1, -x_2, -x_3, -x_4) \cdot (x_1, x_2, x_3, x_4)\end{aligned}$$

so each nonzero 4-tuple is invertible.

Define **H** to be $\mathbf{R}^4$ equipped with $+$ and $\cdot$ as above and call its elements *quaternions*. We must only check associativity to see that **H** is a division ring. There are two approaches that avoid direct computation. First, we assign certain key quaternions their standard notations: $i = (0, 1, 0, 0)$, $j = (0, 0, 1, 0)$, $k = (0, 0, 0, 1)$, and the unit $\mathbf{1} = (1, 0, 0, 0)$. Since $\{\mathbf{1}, i, j, k\}$ is a basis of $\mathbf{R}^4$, any quaternion is a linear combination of these: $q = x_1 \mathbf{1} + x_2 i + x_3 j + x_4 k$. Note that $G = \{\pm \mathbf{1}, \pm i, \pm j, \pm k\}$ is closed under multiplication:

$$i^2 = j^2 = k^2 = -\mathbf{1}, \quad ij = k = -ji, \quad jk = i = -kj, \quad ik = -j = -ki.$$

---

[2] After their discoverer W. R. Hamilton (1805-1865).

We could make a tedious check of 27 cases that G is a group and deduce an associative multiplication on **H** from this (Exercise 7.15).

Figure 7.1. Diagram for quaternionic multiplication.

A second instructive method for establishing associativity on the quaternions uses 2×2 matrices of complex numbers. We define a mapping $T: \mathbf{H} \to M_2(\mathbf{C})$: first break down a quaternion $q = x_1\mathbf{1} + x_2 i + x_3 j + x_4 k$ into two complex numbers $\alpha = x_1 + x_2\sqrt{-1}$ and $\beta = x_3 + x_4\sqrt{-1}$. Define T by

$$T(q) = \begin{pmatrix} x_1 + x_2\sqrt{-1} & x_3 + x_4\sqrt{-1} \\ -x_3 + x_4\sqrt{-1} & x_1 - x_2\sqrt{-1} \end{pmatrix} = \begin{pmatrix} \alpha & \beta \\ -\bar{\beta} & \bar{\alpha} \end{pmatrix}.$$

We note that:

T is clearly an *injective mapping*, since $q_1 \neq q_2$ implies $T(q_1) \neq T(q_2)$. (Recall that quaternions and matrices are unequal if one pair of corresponding coordinates are unequal.)

T is *multiplicative*, i.e. $T(qw) = T(q)T(w)$ for all $q, w \in \mathbf{H}$, since taking q above and letting $w = y_1\mathbf{1} + y_2 i + y_3 j + y_4 k$, $\gamma = y_1 + y_2\sqrt{-1}$, and $\delta = y_3 + y_4\sqrt{-1}$, we compute

$$T(q)T(w) = \begin{pmatrix} \alpha & \beta \\ -\bar{\beta} & \bar{\alpha} \end{pmatrix} \begin{pmatrix} \gamma & \delta \\ -\bar{\delta} & \bar{\gamma} \end{pmatrix} = \begin{pmatrix} \alpha\gamma - \beta\bar{\delta} & \alpha\delta + \beta\bar{\gamma} \\ -\bar{\beta}\gamma - \bar{\alpha}\bar{\delta} & -\bar{\beta}\delta + \bar{\alpha}\bar{\gamma} \end{pmatrix}. \tag{7.2}$$

Now page back to Formula 7.1 where the quaternion product qw is given. Break the quaternion qw into two complex numbers ρ and σ, i.e. $T(qw) = \begin{pmatrix} \rho & \sigma \\ -\bar{\sigma} & \bar{\rho} \end{pmatrix}$. It is easily checked by a computation with complex numbers that $\rho = \alpha\gamma - \beta\bar{\delta}$ and $\sigma = \alpha\delta + \beta\bar{\gamma}$ (cf. Exercise 7.16). Hence T is multiplicative.

We must show that $(q_1q_2)q_3 = q_1(q_2q_3)$ (for all $q_1, q_2, q_3 \in \mathbf{H}$). Since matrix multiplication is associative, it follows that

$$T((q_1q_2)q_3) = T(q_1q_2)T(q_3) = (T(q_1)(T(q_2))T(q_3)$$
$$= T(q_1)(T(q_2)T(q_3)) = T(q_1)T(q_2q_3) = T(q_1(q_2q_3)),$$

so $(q_1q_2)q_3 = q_1(q_2q_3)$ by injectivity of T.[3]

---

[3]There is a principle involved here that could be formulated as follows for sets A and B with one or more binary operations: If $T: A \to B$ is a one-to-one correspondence preserving the binary operations, then whatever laws that hold on B must also hold on A, and vice versa.

7.9. Remark If we identify the space of *pure quaternions*

$$\{a_1 i + a_2 j + a_3 k \mid a_1, a_2, a_3 \in \mathbf{R}\}$$

with the space of 3-vectors $\mathbf{R}^3 = \{v \mid v = v_1 i + v_2 j + v_3 k\}$ equipped with dot product and cross product, the multiplication of quaternions is given by

$$(x_1 \mathbf{1} + v) \cdot (y_1 \mathbf{1} + w) = (x_1 y_1 - v.w)\mathbf{1} + y_1 v + x_1 w + v \times w \qquad (7.3)$$

Check this formula against Formula 7.1 above using your knowledge of vector analysis (Exercise 7.14).[4]

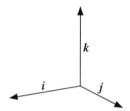

Figure 7.2. The unit vectors in vector analysis.

Certain elements $q \in \mathbf{H}$ *commute* with all other quaternions: $qq' = q'q$ for all $q' \in \mathbf{H}$. For example, $(x\mathbf{1})q = q(x\mathbf{1})$ for all $x \in \mathbf{R}$, by using one of Equations 7.1, 7.3, or 7.2). You will be asked to provide a proof that $\{x\mathbf{1} \mid x \in \mathbf{R}\}$ is the full set of quaternions that commute with all quaternions (Exercise 7.17), and is a field isomorphic to $\mathbf{R}$. This is an example of center, defined below.

7.10. Definition Let F be a division ring. Let $Z(F)$ be the set of $a \in F$ such that $ab = ba$ for all $b \in F$. $Z(F)$ is called the *center* of F.

7.11. Proposition *The center $Z(F)$ of a division ring F is a field.*

Proof Suppose $a, b \in Z(F)$. Then, for all $c \in F$,

$$(a+b)c = ac + bc = ca + cb = c(a+b).$$

Hence, $a + b \in Z(F)$. Also, $ab \in Z(F)$, since

$$(ab)c = a(bc) = a(cb) = \cdots = c(ab)$$

(fill in the missing steps). It follows that $Z(F)$ is closed under addition and multiplication. Also, $(-a)c = -(ac) = -(ca) = c(-a)$, so $-a \in Z(F)$. Moreover, $ab = ba$, so $Z(F)$ is a commutative ring. Finally, if $b \in Z(F) \smallsetminus \{0\}$

$$cb^{-1} = b^{-1}(bc)b^{-1} = b^{-1}(cb)b^{-1} = b^{-1}c \qquad (\forall c \in F).$$

---

[4]Historically, the quaternions led to the cross product on vectors.

Hence $b^{-1} \in Z(F)$. This completes the proof that $Z(F)$ is a field. $\square$

We will find both center and automorphism of division rings crucial to our analytic development of projective geometry in the next chapter.

7.12. Definition An *automorphism* of a division ring F is a one-to-one correspondence $\sigma \colon F \to F$ such that

$$\begin{aligned} \sigma(a+b) &= \sigma(a) + \sigma(b) \\ \sigma(ab) &= \sigma(a)\sigma(b) \end{aligned} \qquad (\forall a, b \in F).$$

It follows that $\sigma(0) = 0$ and $\sigma(1) = 1$. The automorphisms of F form a group under composition, which we denote by $\mathrm{Aut}\, F$.

7.13. Example In Exercise 7.11 you are invited to show that $\sigma_\lambda(x) = \lambda x \lambda^{-1}$ defines an automorphism of F for any $\lambda \neq 0$, called an *inner automorphism*. If $\lambda \in Z(F)$, then $\sigma_\lambda = \mathrm{id}_F$. If $F = \mathbf{H}$, it can be shown that every automorphism is an inner automorphism.

7.3 A Noncommutative Division Ring with Characteristic p

The quaternions and the following division ring will be used as sources of examples in investigating the independence of axioms P1–P4, P5, P6, and P7 in Chapter 8.

7.14. Proposition *There exists a noncommutative division ring of arbitrary characteristic p.*

Proof We sketch the construction of an example due to Hilbert. Certain details will be left to the exercises.

Given a field F with automorphism σ, we show how to form the *skew Laurent series ring in one indeterminate* $F((z; \sigma)) = D$. Elements in D are formal sums with possibly infinitely many nonzero coefficients $a_j \in F$ having a lower bound in index:

$$\frac{a_{-m}}{z^m} + \frac{a_{-(m-1)}}{z^{m-1}} + \cdots + \frac{a_{-1}}{z} + a_0 z^0 + a_1 z + \cdots + a_n z^n + \cdots$$

or equivalently $\sum_{i=-m}^{\infty} a_i z^i$. The lower bound $-m$ will vary from element to element, and might be any integer. We can of course write $\sum_{i=-\infty}^{\infty} a_i z^i$ instead of $\sum_{i=-m}^{\infty} a_i z^i$ by assigning $a_i = 0$ to each $i < -m$. Define an addition on D by

$$\sum_{i=-\infty}^{\infty} a_i z^i + \sum_{i=-\infty}^{\infty} b_i z^i = \sum_{i=-\infty}^{\infty} (a_i + b_i) z^i.$$

Define the multiplication on D by

$$\sum_{i=-n}^{\infty} a_i z^i \sum_{j=-m}^{\infty} b_j z^j = \sum_{k=-n-m}^{\infty} c_k z^k$$

where

$$c_k = \sum_{i+j=k} a_i \sigma^i(b_j).$$

The multiplication is arrived at in three steps:

(1) Enforcing the distributive law.
(2) $z^i b = \sigma^i(b) z^i$. (Here i is zero, positive or negative, so σ^i stands for i successive applications of σ or its inverse, while $\sigma^0 = $ id.)
(3) $z^i z^j = z^{i+j}$.

The multiplication is associative, since

$$(a_i z^i b_j z^j) c_k z^k = a_i \sigma^i(b_j) z^{i+j} c_k z^k = a_i \sigma^i(b_j) \sigma^{i+j}(c_k) z^{i+j+k}$$

$$= a_i \sigma^i(b_j \sigma^j(c_k)) z^i z^{j+k} = a_i z^i (b_j z^j c_k z^k)$$

It is now rather easy to see that D is a ring (Exercise 7.19).

The extraordinary fact is that D is a division ring. Since the unit element is z^0, each z^i is invertible. By Exercise 4.2, it suffices to show that an arbitrary nonzero *Laurent series* $\sum_{i=-m}^{\infty} a_i z^i = f_1$, with $a_{-m} \neq 0$, has a right inverse. Since z^m is invertible, it suffices to show how to find a right inverse of $f = f_1 z^m = \sum_{i=0}^{\infty} b_i z^i$ where $b_i = a_{i-m}$. Suppose $g = \sum_{k=0}^{\infty} c_k z^k$ satisfies $fg = 1$, then

$$b_0 c_0 = 1$$
$$b_1 \sigma(c_0) + b_0 c_1 = 0$$
$$b_2 \sigma^2(c_0) + b_1 \sigma(c_1) + b_0 c_2 = 0$$
$$\vdots$$

Taking $c_0 = b_0^{-1}$, $c_1 = -b_0^{-1} b_1 \sigma(b_0^{-1})$, and defining c_n inductively in terms of the first $n-1$ coefficients of f, it is clear that g above satisfies $fg = 1$. (As an example the reader might recall the formula for the sum of a geometric series $\frac{1}{1-z} = \sum_{k=0}^{\infty} z^k$.)

So far we have built a division ring D for each field F and automorphism $\sigma \colon F \to F$. D will only be noncommutative if we can find a $\sigma \neq$ id. In Exercise 7.11 you will show that taking $F = \mathbf{Z}_p$ won't do for completing the proof since $\mathbf{Z}_p$ has no nontrivial automorphisms. We must therefore build a bigger field over $\mathbf{Z}_p$. One idea is to let $\sigma = $ id and $F = \mathbf{Z}_p$ in our general construction $F((z;\mathrm{id})) = \mathbf{Z}_p((z))$, which is called the Laurent series ring in one indeterminate over $\mathbf{Z}_p$.

Now within the ring $\mathbf{Z}_p((z))$ restrict attention to the subring $\mathbf{Z}_p[z]$ of *polynomials*, elements of the form $\sum_{i=0}^{n} a_i z^i$, having no negative powers of z and whose nonnegative coefficients end at a power n of z (called the *degree* of the polynomial). Now each nonzero polynomial has an inverse in $\mathbf{Z}_p((z))$, so we might consider the set

$$\mathbf{Z}_p(z) = \{ fg^{-1} \mid f, g \in \mathbf{Z}_p[z], g \neq 0 \}.$$

Indeed $\mathbf{Z}_p(z)$ is a subfield of $\mathbf{Z}_p((z))$, called the *field of rational functions in one indeterminate* over $\mathbf{Z}_p$. (See Exercise 7.18 for another, equivalent construction of any field of rational functions in one indeterminate.) On the field $\mathbf{Z}_p(z)$ induce an automorphism σ by sending $z \mapsto z^{-1}$. Thus $\sigma\colon z^2 \mapsto z^{-2}$,

$$p(z) = \sum_{i=0}^{n} a_i z^i \mapsto p(z^{-1}) = a_0 + \sum_{i=1}^{n} a_i z^{-i}.$$

Since $\sigma^2 = \mathrm{id}$, σ is a bijection; it is clear that σ is linear and multiplicative, whence an automorphism.

Now consider $F((x;\sigma))$ where F is the field $\mathbf{Z}_p(z)$, and σ is the automorphism induced by $z \mapsto z^{-1}$. Written side by side, $\mathbf{Z}_p(z)((x;\sigma))$ is a two variable skew Laurent series, but this needn't bother us. We have shown that $\mathbf{Z}_p(z)((x;\sigma))$ is a noncommutative division ring. Note (carefully) that

$$\overbrace{1 + \cdots + 1}^{p \text{ times}} = 0$$

in this ring. Hence we have constructed a noncommutative division ring of characteristic p.   □

Exercises

7.1   Let R be a ring. Show that
   (a) $a0 = 0a = 0$ for all $a \in R$.
   (b) $(-a)c = -(ac) = a(-c)$ for all $a, c \in R$.
   (c) If a is a zero divisor in R, then a has no inverse.

7.2   Suppose R is a ring with no zero divisors, i.e. $ab = 0$ implies $a = 0$ or $b = 0$ for all $a, b \in R$. Then multiplication satisfies left (and right) cancellation: $x \neq 0$ and $xy_1 = xy_2$ implies $y_1 = y_2$. Prove.

7.3   Let $\mathrm{M}_n(R)$ denote the set of $n \times n$ matrices over an arbitrary ring R. Show that $\mathrm{M}_n(R)$ is a ring.

7.4   Check that congruence modulo n is an equivalence relation on $\mathbf{Z}$.

7.5   Check the distributivity of multiplication over addition in $\mathbf{Z}_n$.

7.6   Prove that if a and b are relatively prime, then there exists m and n such that $am + bn = 1$.

7.7   Show that conjugation is an automorphism of $\mathbf{C}$. Why is it not an inner automorphism?

7.8   Define an operation on $\mathbf{H}$ called *conjugation* (or *involution*): this is a mapping $\mathbf{H} \rightarrow \mathbf{H}$ defined by $q \mapsto q^* = x_1 \mathbf{1} - x_2 i - x_3 j - x_4 k$ where $q = x_1 \mathbf{1} + x_2 i + x_3 j + x_4 k$. Show that

   (a) $q^{-1} = q^*/|q|^2$.
   (b) $(q_1 + q_2)^* = q_1^* + q_2^*$.
   (c) $(q_1 q_2)^* = q_2^* q_1^*$.
   (d) **Euler's formula** . Apply (a)–(c) to establish it:

$$(x_1^2 + x_2^2 + x_3^2 + x_4^2)(y_1^2 + y_2^2 + y_3^2 + y_4^2) =$$
$$(x_1 y_1 - x_2 y_2 - x_3 y_3 - x_4 y_4)^2 + (x_1 y_2 + x_2 y_1 + x_3 y_4 - x_4 y_3)^2$$
$$+ (x_1 y_3 + x_2 y_1 + x_4 y_2 - x_2 y_4)^2 + (x_1 y_4 + x_4 y_1 + x_2 y_3 - x_3 y_2)^2.$$

7.9   An *isomorphism* of rings R_1 and R_2 is a one-to-one correspondence $\psi: R_1 \rightarrow R_2$ that satisfies

   (1) $\psi(x+y) = \psi(x) + \psi(y)$,
   (2) $\psi(xy) = \psi(x)\psi(y)$

for all $x, y \in R_1$. Show that
   (a) $\psi(1) = 1$.
   (b) If $u \in R_1$ is invertible, then $\psi(u)$ is invertible in R_2.
   (c) $\mathbf{H}$ is not isomorphic to $M_2(\mathbf{R})$.

7.10   The quaternions are considered to be an extension of the complex numbers and the reals, as seen in part (a) below.
   (a) Show that the subset $\{x\mathbf{1} + yi \mid x, y \in \mathbf{R}\}$ of $\mathbf{H}$ forms a subring isomorphic to $\mathbf{C}$. Show the same for $\{x\mathbf{1} \mid x \in \mathbf{R}\}$ and $\mathbf{R}$.
   (b) Show that $X^2 + Y^2$ can be factored over the complex numbers. Show that $X^2 + Y^2 + Z^2$ can be factored over the quaternions.
   (c) Show that $X^2 + 1$ has no roots in $\mathbf{R}$, two roots in $\mathbf{C}$, and an infinite number of roots in $\mathbf{H}$.

7.11   (a) Let F be a division ring, and let λ be a fixed nonzero element of F. Prove that the map $\phi: F \rightarrow F$, defined by $\phi(x) = \lambda x \lambda^{-1}$ for all $x \in F$, is an automorphism of F.
   (b) Let p be a prime number. Prove that the field $\mathbf{Z}_p$ of p elements has no automorphisms other than the identity.

7.12   Let $k = \{0, 1, 2\}$ be the field of 3 elements, with addition and multiplication modulo 3. Let $F = \{a + bj \mid a, b \in k\}$, where j is a symbol.
   (a) Define addition and multiplication in F, using the relation $j^2 = 2$. Check that F is then a field.

(*b*) Prove that the multiplicative group F^* of nonzero elements of F is cyclic of order 8.

(*c*) Find a nontrivial automorphism of F.

7.13 A commutative ring R with no zero divisors is called an *integral domain*. Define its *field of fractions* F and embed R in F as follows.

(*a*) On $R \times (R \smallsetminus \{0\})$, introduce the relation $(a,b) \sim (c,d)$ if $ad = bc$. Show that $\sim$ is an equivalence relation. Let $\frac{a}{b}$ denote the equivalence class of (a,b).

(*b*) Define addition by $\frac{a}{b} + \frac{c}{d} = \frac{ad+bc}{bd}$. Check that this formula is not dependent on choice of representative $\frac{a}{b}$ or $\frac{c}{d}$. Show that $F = \{ \frac{a}{b} \mid a \in R, b \in R \smallsetminus \{0\} \}$ is an abelian group under $+$.

(*c*) Define multiplication by $\frac{a}{b} \cdot \frac{c}{d} = \frac{ac}{bd}$ and check that $(F, +, \cdot)$ is a field. If R is the ring of integers, which is its field of fractions?

(*d*) Show that the mapping $R \to F$ given by $x \mapsto \frac{x}{1}$ is a ring homomorphism.

(*e*) If $\phi: R \to F_1$ is a ring homomorphism of R into a field F_1 and ι is the mapping defined in (*d*), show that there is a uniquely determined ring homomorphism $\alpha: F \to F_1$ such that $\phi = \alpha \circ \iota$.

7.14 Define on $\mathbf{R} \times \mathbf{R}^3 = \{ (x,v) \mid x \in \mathbf{R}, v \in \mathbf{R}^3 \}$ an addition $(x,v) + (y,w) = (x+y, v+w)$ and multiplication

$$(x,v) \cdot (y,w) = (xy - v.w, xw + yv + v \times w)$$

where we have used vector addition, dot product, cross product, and scalar multiplication familiar from vector analysis.

Let $\{i, j, k\}$ denote the standard basis of orthogonal unit vectors in $\mathbf{R}^3$. Show that the mapping $\psi: \mathbf{H} \to \mathbf{R} \times \mathbf{R}^3$ given by

$$x_1 \mathbf{1} + x_2 i + x_3 j + x_4 k \mapsto (x_1, x_2 i + x_3 j + x_4 k)$$

is a one-to-one correspondence that is linear, i.e., $\psi(x+y) = \psi(x) + \psi(y)$ for all $x, y \in \mathbf{H}$, and multiplicative, i.e., $\psi(xy) = \psi(x)\psi(y)$. Using the fact that $\mathbf{H}$ is a division ring, show that $\mathbf{R} \times \mathbf{R}^3$ is a division ring isomorphic to $\mathbf{H}$. (You should establish symbolically that ψ carries associativity, etc., over to $\mathbf{R} \times \mathbf{R}^3$. What is the principle applicable to sets with one or more binary operations?)

7.15 Verify the associative law on the set $G = \{ \pm 1, \pm i, \pm j, \pm k \}$ of quaternions using formula 7.1. Show that G is a nonabelian group.

7.16 (*a*) Check that the subset

$$D = \left\{ \begin{pmatrix} \alpha & \beta \\ -\bar{\beta} & \bar{\alpha} \end{pmatrix} \middle| \alpha, \beta \in \mathbf{C} \right\}$$

of $M_2(\mathbf{C})$ is a division ring.

(*b*) Show that the mapping $T: \mathbf{H} \to D$ given by

$$T(x_1 \mathbf{1} + x_2 i + x_3 j + x_4 k) = \begin{pmatrix} x_1 + x_2\sqrt{-1} & x_3 + x_4\sqrt{-1} \\ -x_3 + x_4\sqrt{-1} & x_1 - x_2\sqrt{-1} \end{pmatrix}$$

is a ring isomorphism.

    (c) Show that $\det T(q) = |q|$ and $T(q^*) = \overline{T(q)}^T$ where $\bar{A}^T$ denotes the complex transpose of matrix A.

7.17   Prove that the center of $\mathbf{H}$ is isomorphic to the field $\mathbf{R}$.

7.18   Let F be a field. Define the ring of *polynomials* in one indeterminate $F[x]$ as follows.
    (a) Let $x^0, x^1, x^2, \ldots, x^n, \ldots$ denote the basis elements of an infinite dimensional vector space V over F. Let $F[x]$ be the set of all finite linear combinations of these denoted by $\sum_{i=0}^{n} a_i x^i$, where either $a_n \neq 0$ or $n = 0$, $a_0 = 0$. An element $a_0 x^0 + \cdots + a_n x^n$ in $F[x]$ is called a polynomial of *degree n*. Show that $F[x]$ inherits an addition from V such that $F[x]$ is an abelian group and $\deg(f+g) \leq \max\{\deg(f), \deg(g)\}$, where $\deg(f)$ denotes the degree of the polynomial f.
    (b) Define a multiplication first on the basis elements: $x^i x^j = x^{i+j}$. Extend this multiplication to $F[x]$ distributively and letting coefficients commute past the powers of x. Show that this gives

$$\sum_{i=0}^{m} a_i x^i \sum_{j=0}^{n} b_j x^j = \sum_{k=0}^{m+n} c_k x^k$$

where $c_k = \sum_{i=0}^{k} a_i b_{k-i}$, with $c_0 = a_0 b_0$, $c_1 = a_0 b_1 + a_1 b_0, \ldots$. Show that $\deg(fg) = \deg(f) + \deg(g)$ for all $f, g \in F[x]$.
    (c) Show that $F[x]$ is an integral domain and identifiable via an isomorphism with the polynomials in the Laurent series ring $F((x; \mathrm{id})) = F((x))$. Applying Exercise 7.13, its field of fractions is denoted by $F(x)$ and called the *field of rational functions*. Show that the field of fractions $F(x)$ is isomorphic with the subfield $\{fg^{-1} \mid f, g \in F[x], g \neq 0\}$ in the field $F((x))$.

7.19   Use an infinite vector space with basis as in Exercise 7.18 to define the skew Laurent series ring $F((z; \sigma))$ over field F and automorphism $\sigma \colon F \to F$ (cf. Proposition 7.14). Prove that $F((z; \sigma))$ is a ring without zero divisors.

7.20   Given a ring R, define $U(R) = \{x \in R \mid xy = 1 = yx \text{ some } y \in R\}$. Show that $U(R)$ is a group, the *group of units* of R.

7.21   Let σ be an automorphism of the division ring R, and x a nonzero number in R. Show that $\sigma(x^{-1}) = \sigma(x)^{-1}$.

7.22   **Hua's identity.** Let a and b be elements of a division ring such that $a \neq b^{-1}$. Show that the following identity holds:

$$((a - b^{-1})^{-1} - a^{-1})^{-1} = aba - a.$$

Chapter 8

Projective Planes over Division Rings

In this chapter, we introduce the projective plane over a division ring. This will give us many examples of projective planes aside from the ones we know already. We will determine the automorphism groups of these projective planes. Next, we will study various geometric properties of the projective plane corresponding to algebraic properties of the division ring. Finally, we use our many new examples of projective planes to investigate the independence of the axioms P1–P4, P5–P7.

8.1 $\mathbf{P}^2(R)$

A division ring has all the properties of the real numbers, except for commutativity of multiplication, ordering and completeness. Now we define the projective plane over a division ring, mimicking the analytic definition of the real projective plane (Section 2.2). What is needed to successfully meet the conditions P1–P4 (and indeed P5) is just the possibility of doing the four basic operations, addition, subtraction, multiplication, and division.

8.1. Definition Let R be a division ring. We define the *projective plane over R*, written $\mathbf{P}^2(R)$, as follows. A point of the projective plane is an equivalence class of triples $P = (x_1, x_2, x_3)$ where $x_1, x_2, x_3 \in R$ are not all zero, and where the two triples are equivalent,

$$(x_1, x_2, x_3) \sim (x_1', x_2', x_3'),$$

if and only if there is an element $\lambda \in R$, $\lambda \neq 0$, such that

$$x_i' = x_i \lambda \qquad \text{for } i = 1, 2, 3.$$

87

(Note that we multiply by λ on the *right*. It is important to keep this in mind, since the multiplication may not be commutative.)

A line in $\mathbf{P}^2(R)$ is the set of all points satisfying a linear equation of the form

$$c_1 x_1 + c_2 x_2 + c_3 x_3 = 0$$

where $c_1, c_2, c_3 \in R$ are not all zero. Note that we multiply here on the left, so that this equation actually has equivalence classes of triples instead of triples in its solution set.

Now one can check that the axioms P1–P4 are satisfied, and so $\mathbf{P}^2(R)$ is a projective plane (Exercise 8.1).

8.2. Example If $R = \mathbf{Z}_2$ is the field of two elements, then $\mathbf{P}^2(R)$ is the projective plane of seven points.

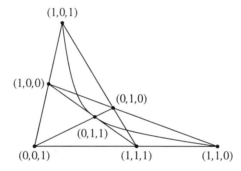

Figure 8.1. An isomorphism of $\mathbf{P}^2(\mathbf{Z}_2)$ with π_7.

8.3. Example More generally, if $R = \mathbf{Z}_p$ for any prime number p, then $\mathbf{P}^2(R)$ is a projective plane with $p^2 + p + 1$ points. Indeed, any line has $p + 1$ points, so this follows from Exercise 2.4.

8.4. Example If $R = \mathbf{R}$ we get back the real projective plane.

8.5. Theorem *The plane* $\mathbf{P}^2(R)$ *over a division ring always satisfies Desargues' axiom P5.*

Proof One could argue the same way as in the proof of Theorem 3.1 if R is a field. For the general case we apply Theorem 3.6. One defines projective 3-space $\mathbf{P}^3(R)$ by taking points to be equivalence classes (x_1, x_2, x_3, x_4), $x_i \in R$ not all zero, and where this is equivalent of $(x_1\lambda, x_2\lambda, x_3\lambda, x_4\lambda)$. Planes are defined by left linear equations, $\sum_{i=1}^4 c_i x_i = 0$, and lines as intersections of distinct planes. Now it is an exercise to check that the axioms S1–S6 are satisfied, so $\mathbf{P}^3(R)$ is a projective space (Exercise 8.9).

Then $\mathbf{P}^2(R)$ is embedded as the plane $x_4 = 0$ in this projective 3-space, and so P5 holds there by Theorem 3.6.   □

8.6. N-dimensional projective geometry Define $\mathbf{P}^n(R)$, n-dimensional projective space over an arbitrary division ring R, as the set of points $(x_1,\dots,x_{n+1})$ in R^{n+1} subject to the equivalence relation

$$(x_1,\dots,x_{n+1}) \sim (x_1,\dots,x_{n+1})\lambda = (x_1\lambda,\dots,x_{n+1}\lambda).$$

n-planes, or *hyperplanes*, are the points satisfying left linear equations like

$$\sum_{i=1}^{n+1} c_i x_i = 0.$$

k-planes are sets of points satisfying simultaneously a system of $n+1-k$ different left linear equations. The 1-planes are points, and 2-planes are lines. Axioms for n-space are not hard to give: see [Seidenberg]. One can also work analytically with $\mathbf{P}^n(R)$: see [Samuel] or [Yale]. It is a relatively easy project to generalize all the results in the next section to $\mathrm{Aut}\,\mathbf{P}^n(R)$, the group of bijections of $\mathbf{P}^n(R)$ into itself which send k-planes into k-planes.

8.2   The Automorphism Group of $\mathbf{P}^2(R)$

In this section, we let R be a division ring. We will study the group $\mathrm{Aut}\,\mathbf{P}^2(R)$ of automorphisms of our projective plane. We will show that the automorphisms of $\mathbf{P}^2(R)$ consist of *semi-linear transformations* T which satisfy

$$T(x+y) = T(x) + T(y) \qquad (x, y \in R^3)$$

and for which there exists a ring automorphism $\alpha\colon R \to R$ such that

$$T(\lambda x) = \alpha(\lambda)T(x) \qquad (\lambda \in R).$$

(It may be helpful to read this section first to see what it says for the real numbers, and secondly for what it says in the general case of division rings.)

8.7. Definition An $n \times n$ matrix $A = (a_{ij})$ of elements of R is *invertible* if there is an $n \times n$ matrix B, such that $AB = BA = I$, the identity matrix. B is called the inverse of A and denoted by A^{-1}.

    If we are working over a field F, the invertible matrices are just the matrices with determinant $\neq 0$. Over general division rings, determinants do not make sense.

8.8. Proposition *Let $A = (a_{ij})$ be an invertible 3×3 matrix of elements of R.*
Then the equations

$$x_i' = \sum_{j=1}^{3} a_{ij}x_j \qquad (i = 1,2,3)$$

define an automorphism T_A of $\mathbf{P}^2(R)$ given by $(x_1,x_2,x_3) \mapsto (x_1',x_2',x_3')$.

Proof We must observe several things.

(*1*) If we replace (x_1,x_2,x_3) by $(x_1\lambda,x_2\lambda,x_3\lambda)$, then, by the right distributive law, (x_1',x_2',x_3') is replaced by $(x_1'\lambda,x_2'\lambda,x_3'\lambda)$, so the mapping is well-
defined. We must also check that x_1',x_2',x_3' are not all zero. Indeed, in matrix
notation

$$\begin{pmatrix} a_{11} & a_{12} & a_{13} \\ a_{21} & a_{22} & a_{23} \\ a_{31} & a_{32} & a_{33} \end{pmatrix} \begin{pmatrix} x_1 \\ x_2 \\ x_3 \end{pmatrix} = \begin{pmatrix} x_1' \\ x_2' \\ x_3' \end{pmatrix}$$

or in compact notation, $Ax = x'$. But since A has inverse A^{-1}, we can multiply
on the left by A^{-1}. We get $x = A^{-1}x'$ by an appeal to associativity of the
ring $M_n(R)$ of square matrices over R (Exercise 7.3: let a column vector $\begin{pmatrix} y_1 \\ y_2 \\ y_3 \end{pmatrix}$
stand for the matrix $\begin{pmatrix} y_1 & 0 & 0 \\ y_2 & 0 & 0 \\ y_3 & 0 & 0 \end{pmatrix}$). So if the x_i' are all zero, the x_i are also all zero,
which is not possible for a point of $\mathbf{P}^2(R)$. Thus T_A is a well-defined map of
$\mathbf{P}^2(R)$ into $\mathbf{P}^2(R)$.

(*2*) The expression $x = A^{-1}x'$ shows that $T_{A^{-1}}$ is the inverse mapping to
T_A, hence T_A must be one-to-one and surjective.

(*3*) Finally, we must check that T_A takes lines into lines. Indeed, let

$$c_1x_1 + c_2x_2 + c_3x_3 = 0 \qquad (8.1)$$

be the equation of a line. We must find a new line, so that whenever (x_1,x_2,x_3)
satisfies Equation 8.1, its image (x_1',x_2',x_3') lies on the new line. Let $A^{-1} =$
(b_{ij}). Then we have $x_i = \sum_{j=1}^{3} b_{ij}x_j'$ for each i. Thus if (x_1,x_2,x_3) satisfies
Equation (8.1), then (x_1',x_2',x_3') will satisfy the equation

$$c_1 \sum_{j=1}^{3} b_{1j}x_j' + c_2 \sum_{j=1}^{3} b_{2j}x_j' + c_3 \sum_{j=1}^{3} b_{3j}x_j' = 0$$

which is

$$\left(\sum_{i=1}^{3} c_i b_{i1} \right) x_1' + \left(\sum_{i=1}^{3} c_i b_{i2} \right) x_2' + \left(\sum_{i=1}^{3} c_i b_{i3} \right) x_3' = 0.$$

This is the equation of the required line. We have only to check that the three coefficients

$$c'_j = \sum_{i=1}^{3} c_i b_{ij} \qquad (j = 1, 2, 3) \tag{8.2}$$

are not all zero. But this argument is analogous to the argument in (*1*) above: Equation 8.2 represents the fact that

$$(c_1, c_2, c_3) \cdot A^{-1} = (c'_1, c'_2, c'_3)$$

where $(c_1, c_2, c_3) = \begin{pmatrix} c_1 & c_2 & c_3 \\ 0 & 0 & 0 \\ 0 & 0 & 0 \end{pmatrix}$. Multiplying by A on the right shows that the c_i can be expressed in terms of the c'_i. Hence if the c'_i were all zero, the c_i would all be zero, which contradicts the definition of line in $\mathbf{P}^2(R)$.

Hence T_A is an automorphism of $\mathbf{P}^2(R)$. □

8.9. Lemma *Let A and A' be two invertible matrices. Then T_A and $T_{A'}$ have the same effect on the four points $P_1 = (1, 0, 0)$, $P_2 = (0, 1, 0)$, $P_3 = (0, 0, 1)$, and $P_4 = (1, 1, 1)$ if and only if there is a $\mu \in R$, $\mu \neq 0$, such that $A' = A\mu$.*

Proof Clearly, if there is such a μ, $T_A(P_i) = T_{A'}(P_i)$ for $i = 1, 2, 3, 4$, by a direct computation.

Conversely, suppose $T_A = T_{A'}$. We will then study the action of T_A and $T_{A'}$ on four specific points of $\mathbf{P}^2(R)$, namely $(1, 0, 0)$, $(0, 1, 0)$, $(0, 0, 1)$, and $(1, 1, 1)$, i.e. P_1, P_2, P_3, and P_4, respectively. Give A and A' the usual coefficients a_{ij} and a'_{ij}, respectively. Now

$$T_A(P_1) = \begin{pmatrix} a_{11} & a_{12} & a_{13} \\ a_{21} & a_{22} & a_{23} \\ a_{31} & a_{32} & a_{33} \end{pmatrix} \begin{pmatrix} 1 \\ 0 \\ 0 \end{pmatrix} = \begin{pmatrix} a_{11} \\ a_{21} \\ a_{31} \end{pmatrix}$$

and

$$T_{A'}(P_1) = A' \cdot \begin{pmatrix} 1 \\ 0 \\ 0 \end{pmatrix} = \begin{pmatrix} a'_{11} \\ a'_{21} \\ a'_{31} \end{pmatrix}.$$

Now these two sets of coordinates are supposed to represent the same points of $\mathbf{P}^2(R)$, so there must exist a $\lambda_1 \in R$, $\lambda_1 \neq 0$, such that

$$a'_{11} = a_{11}\lambda_1,$$
$$a'_{21} = a_{21}\lambda_1,$$
$$a'_{31} = a_{31}\lambda_1.$$

Similarly, applying T_A and $T_{A'}$, to the points P_2 and P_3, we find the numbers $\lambda_2 \in R$ and $\lambda_3 \in R$, both $\neq 0$, such that

$$
\begin{aligned}
a'_{12} &= a_{12}\lambda_2 & a'_{13} &= a_{13}\lambda_3 \\
a'_{22} &= a_{22}\lambda_2 & a'_{23} &= a_{23}\lambda_3 \\
a'_{32} &= a_{32}\lambda_2 & a'_{33} &= a_{33}\lambda_3
\end{aligned}
$$

Now apply T_A and $T_{A'}$ to the point P_4. Since there is a $\mu \in R \smallsetminus \{0\}$ such that $T_{A'}(P_4) = T_A(P_4)\mu$, it follows that

$$
A \cdot \begin{pmatrix} 1 \\ 1 \\ 1 \end{pmatrix} \mu = \begin{pmatrix} a_{11}+a_{12}+a_{13} \\ a_{21}+a_{22}+a_{23} \\ a_{31}+a_{32}+a_{33} \end{pmatrix} \mu = \begin{pmatrix} a'_{11}+a'_{12}+a'_{13} \\ a'_{21}+a'_{22}+a'_{23} \\ a'_{31}+a'_{32}+a'_{33} \end{pmatrix}
$$
$$
= \begin{pmatrix} a_{11}\lambda_1+a_{12}\lambda_2+a_{13}\lambda_3 \\ a_{21}\lambda_1+a_{22}\lambda_2+a_{23}\lambda_3 \\ a_{31}\lambda_1+a_{32}\lambda_2+a_{33}\lambda_3 \end{pmatrix}
$$

Now applying subtraction to the equation above, we find that

$$
\begin{aligned}
a_{11}(\lambda_1 - \mu) + a_{12}(\lambda_2 - \mu) + a_{13}(\lambda_3 - \mu) &= 0 \\
a_{21}(\lambda_1 - \mu) + a_{22}(\lambda_2 - \mu) + a_{23}(\lambda_3 - \mu) &= 0 \\
a_{31}(\lambda_1 - \mu) + a_{32}(\lambda_2 - \mu) + a_{33}(\lambda_3 - \mu) &= 0.
\end{aligned}
$$

In other words, the vector $(\lambda_1 - \mu, \lambda_2 - \mu, \lambda_3 - \mu)$ is sent into $(0,0,0)$ under the linear mapping A. Hence $\lambda_1 = \lambda_2 = \lambda_3 = \mu$. (We saw this in the proof of Proposition 8.2: an ordered triple of numbers, not all zero, cannot be sent into $(0,0,0)$ by A. Hence $\lambda_1 - \mu = 0$, $\lambda_2 - \mu = 0$, and $\lambda_3 - \mu = 0$.)

So $A' = A\mu$, and we are done. $\square$

8.10. Lemma *Let $\lambda \in R$, $\lambda \neq 0$, and consider the diagonal matrix λI. Then $T_{\lambda I}$ is the identity transformation of $\mathbf{P}^2(R)$ if and only if λ is in the center of R. Otherwise, $T_{\lambda I}$ is the automorphism given by $(x_1, x_2, x_3) \mapsto (\sigma(x_1), \sigma(x_2), \sigma(x_3))$ where σ is the automorphism of R given by $x \mapsto \lambda x \lambda^{-1}$. I.e. σ is an inner automorphism.*

Proof For any $\lambda \neq 0$, $T_{\lambda I}$ sends (x_1, x_2, x_3) into the point $(\lambda x_1, \lambda x_2, \lambda x_3)$, which also has the homogeneous coordinates $(\lambda x_1 \lambda^{-1}, \lambda x_2 \lambda^{-1}, \lambda x_3 \lambda^{-1})$. Thus $T_{\lambda I}$ is induced by an inner automorphism of the division ring, as stated.

Let $x_1 = x$, $x_2 = 1$, $x_3 = 1$. Then it is clear that $T_{\lambda I}$ is the identity automorphism of $\mathbf{P}^2(R)$ iff

$$
\begin{pmatrix} \lambda x \\ \lambda \\ \lambda \end{pmatrix} = T_{\lambda I} \begin{pmatrix} x \\ 1 \\ 1 \end{pmatrix} = \begin{pmatrix} x \\ 1 \\ 1 \end{pmatrix} \lambda = \begin{pmatrix} x\lambda \\ \lambda \\ \lambda \end{pmatrix},
$$

hence iff $\lambda x = x\lambda$ for all x, i.e., λ is in the center of R. $\square$

8.11. Definition We denote by $\mathrm{PGL}(2,R)$ the group of automorphisms of $\mathbf{P}^2(R)$ of the form T_A for some invertible 3×3 matrix A. $\mathrm{PGL}(2,R)$ stands for the projective general linear group of the plane over R. (Note that if B is another invertible 3×3 matrix, then $T_{AB} = T_A T_B$. Also, $T_I = \mathrm{id}$ and $T_A^{-1} = T_{A^{-1}}$, so that $\mathrm{PGL}(2,R)$ is indeed a group.)

8.12. Proposition *Let A and A' be invertible matrices. Then $T_A = T_{A'}$ if and only if there exists a nonzero λ in the center of R such that $A' = A\lambda$.*

Proof The "only if" part is ok since $T_{A'}(x) = A\lambda(x) = A(x)\lambda = T_A(x)$ for every projective point x. Conversely, if $T_A = T_{A'}$, then by Lemma 8.9, $A' = A\lambda = A \cdot (\lambda I)$. So $T_{A'} = T_{A \cdot \lambda I} = T_A \circ T_{\lambda I}$, whence $T_{\lambda I}$ is the identity. By Lemma 8.10, λ lies in the center of R. $\qquad\square$

The next theorem is a fundamental theorem in projective geometry. It says the automorphisms of $\mathbf{P}^2(R)$ are transitive on ordered complete quadrangles. You may apply this theorem to find the order of the automorphism groups of the finite projective planes (Exercise 8.12).

8.13. Theorem *Let A,B,C,D and A',B',C',D' be two quadruples of points, no 3 collinear. Then there is an automorphism $T \in \mathrm{PGL}(2,R)$ such that $T(A) = A'$, $T(B) = B'$, $T(C) = C'$, and $T(D) = D'$. If R is a field, then T is unique.*

Proof Let P_1, P_2, P_3, P_4 be the four points $(1,0,0)$, $(0,1,0)$, $(0,0,1)$, $(1,1,1)$ considered above. Then it will be sufficient to prove the theorem in the case $A,B,C,D = P_1, P_2, P_3, P_4$. Indeed, the quadruple P_1, P_2, P_3, P_4 can be sent into any other. Let ϕ send it to A,B,C,D, and let ψ send it to A',B',C',D'. Then $\psi\phi^{-1}$ sends A,B,C,D into A',B',C',D'.

Let A,B,C,D have the homogeneous coordinates (a_1,a_2,a_3), (b_1,b_2,b_3), (c_1,c_2,c_3), (d_1,d_2,d_3), respectively. Then we must find an invertible 3×3 matrix (t_{ij}) and numbers λ,μ,ν,ρ such that

$$
\begin{array}{lllll}
T(P_1) = A & \Longleftrightarrow & a_i\lambda = t_{i1} & (i = 1,2,3) \\
T(P_2) = B & \Longleftrightarrow & b_i\mu = t_{i2} & (i = 1,2,3) \\
T(P_3) = C & \Longleftrightarrow & c_i\nu = t_{i3} & (i = 1,2,3) \\
T(P_4) = D & \Longleftrightarrow & d_i\rho = t_{i1} + t_{i2} + t_{i3} & (i = 1,2,3).
\end{array}
$$

Then clearly it will be sufficient to find nonzero λ,μ,ν in R such that

$$
\begin{aligned}
a_1\lambda + b_1\mu + c_1\nu &= d_1 \\
a_2\lambda + b_2\mu + c_2\nu &= d_2 \\
a_3\lambda + b_3\mu + c_3\nu &= d_3
\end{aligned}
\tag{8.3}
$$

8.14. Lemma *Let A, B, C be three points in $\mathbf{P}^2(R)$, with coordinates (a_1, a_2, a_3), (b_1, b_2, b_3), and (c_1, c_2, c_3), respectively. Then A, B, C are noncollinear if and only if $\begin{pmatrix} a_1 & b_1 & c_1 \\ a_2 & b_2 & c_2 \\ a_3 & b_3 & c_3 \end{pmatrix}$ is invertible.*

Proof of lemma The points A, B, C are collinear if and only if there is a line with equation say $h_1 x_1 + h_2 x_2 + h_3 x_3 = 0$, the h_i not all zero, such that this equation is satisfied by the coordinates of A, B, C. We have seen that the matrix $A = (a_{ij})$ is invertible iff for each set of numbers $b = (b_i)$, the set of linear equations corresponding to $Ax = b$ have a unique solution $x = A^{-1}b$. It follows that A is invertible if and only if for $b_i = 0$, the set of equations $\sum a_{ij} x_j = b_i$ has only the trivial solution, $x = 0$ (the "only if" part is left as Exercise 8.4). Now our h_i are solutions of a set of analogous equations in row vector notation. Indeed

$$(h_1, h_2, h_3) \cdot \begin{pmatrix} a_1 & b_1 & c_1 \\ a_2 & b_2 & c_2 \\ a_3 & b_3 & c_3 \end{pmatrix} = (0, 0, 0).$$

Therefore a nontrivial solution (h_1, h_2, h_3) does not exist iff the matrix above is invertible, whence no line contains A, B, and C.

Proof of theorem, continued Now A, B, C are noncollinear by hypothesis. Hence by the lemma, $\begin{pmatrix} a_1 & b_1 & c_1 \\ a_2 & b_2 & c_2 \\ a_3 & b_3 & c_3 \end{pmatrix}$ is invertible. Hence we can solve Equation 8.3 for λ, μ, ν not all zero. Now we claim that λ, μ, ν are all $\neq 0$. Indeed, suppose with no loss of generality that $\lambda = 0$. Then our equations say that

$$b_1 \mu + c_1 \nu - d_1 = 0$$
$$b_2 \mu + c_2 \nu - d_2 = 0$$
$$b_3 \mu + c_3 \nu - d_3 = 0$$

and hence $\begin{pmatrix} b_1 & c_1 & d_1 \\ b_2 & c_2 & d_2 \\ b_3 & c_3 & d_3 \end{pmatrix}$ is noninvertible, which is impossible, by the lemma, since B, C, D are not collinear.

So we have found nonzero λ, μ, ν which satisfy the equations above. We define t_{ij} by the equations $a_i \lambda = t_{i1}$, $b_i \mu = t_{i2}$, and $c_i \nu = t_{i3}$. Then (t_{ij}) is a matrix, which is invertible (again by the lemma, since A, B, C are noncollinear!), so T, given by the matrix (t_{ij}), is an element of $PGL(2, R)$ which sends P_1, P_2, P_3, P_4 to A, B, C, D.

For the uniqueness in the case of a field F, suppose that T and T' are two elements of $PGL(2, F)$ which accomplish our task. Then by Lemma 8.9, $T' = T\mu$ for some $\mu \in F \setminus \{0\}$, and hence give the same element of $PGL(2, F)$. $\square$

Note that in general the transformation T is not unique. For example, over the quaternions $\mathbf{H}$ there are many nontrivial inner automorphisms (Exercise 7.11); each will induce automorphisms fixing P_1, P_2, P_3, and P_4.

8.15. Proposition *Let ϕ be any automorphism of $\mathbf{P}^2(R)$ which leaves fixed the four points P_1, P_2, P_3, P_4 mentioned above. Then there is an automorphism $\sigma \in \operatorname{Aut} R$, such that $\phi(x_1, x_2, x_3) = (\sigma(x_1), \sigma(x_2), \sigma(x_3))$.*

Proof We note that ϕ must leave the line $x_3 = 0$ fixed since it contains P_2 and P_1. We will take this line as the line at infinity, and consider the affine plane $x_3 \neq 0$: $\mathbf{A} = \mathbf{P}^2(R) \smallsetminus \{x_3 = 0\}$. (Refer to Exercise 2.2.)

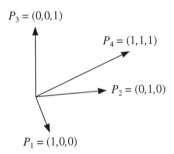

Figure 8.2. The points P_1–P_4 of the standard quadrangle.

Our automorphism ϕ then sends $\mathbf{A}$ into itself, and so is an automorphism of the affine plane. We will use affine coordinates $x = x_1 x_3^{-1}$, $y = x_2 x_3^{-1}$. Then P_1 and P_2 refer to ideal points on lines of slope 0 and ∞, respectively, while P_3 and P_4 receive affine coordinates $(0,0)$ and $(1,1)$, respectively. Since ϕ leaves fixed P_1 and P_2, it will send horizontal lines into horizontal lines, vertical lines into vertical lines. Since ϕ leaves fixed $(0,0)$ and $(1,1)$, it leaves fixed the x-axis and the y-axis.

Let $(a,0)$ be a point on the x-axis. Then $\phi(a,0)$ is also on the x-axis, so it can be written as $(\sigma(a),0)$ for a suitable element $\sigma(a) \in R$. Thus we define a mapping $\sigma: R \to R$, and we see immediately that $\sigma(0) = 0$. Also $\sigma(1) = 1$ since ϕ fixes the horizontal line $y = 0$ and vertical line $x = 1$, therefore their intersection $(1,0)$.

The line $x = y$ is sent into itself, because P_3 and P_4 are fixed. Since vertical lines go into vertical lines, the point

$$(a,a) = (\text{line } x = y) \cap (\text{line } x = a)$$

is sent into

$$(\sigma(a), \sigma(a)) = (\text{line } x = y) \cap (\text{line } x = \sigma(a)).$$

Similarly, horizontal lines go into horizontal lines, and the y-axis goes into itself, so we deduce that $\phi(0,a) = (0,\sigma(a))$. Finally, if (a,b) is any point, we deduce by drawing lines $x = a$ and $y = b$ that $\phi(a,b) = (\sigma(a),\sigma(b))$.

Hence the action of ϕ on the affine plane is completely expressed by the mapping $\sigma: R \to R$ which we have constructed.

Of course, since ϕ is an automorphism of **A**, it must send the x-axis onto itself in a one-to-one manner, so σ is one-to-one and onto.

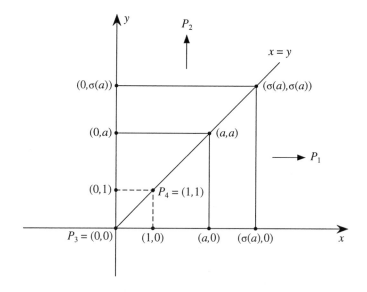

Figure 8.3. Definition of σ.

Now we will show that σ is an automorphism of R. Let $a,b \in R \setminus \{0\}$, and consider the points $(a,0)$, $(b,0)$ on the x-axis. We can construct the point $(a+b,0)$ geometrically as follows:

(1) Draw the line n joining $(0,1)$ and $(b,0)$.

(2) Draw the line r parallel to n through $(a,1)$.

(3) Intersect r with the x-axis. This is the point $(a+b,0)$ since r clearly has equation $y = -b^{-1}x + 1 + b^{-1}a$.

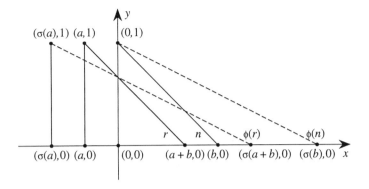

Figure 8.4. $\sigma(a) + \sigma(b) = \sigma(a+b)$.

Now ϕ preserves joins, intersections and parallelism. Then $\phi(r) \parallel \phi(n)$, $\phi(r)$ a line through $(\sigma(a), 1)$, and $\phi(n)$ the line through $(0, 1)$ and $(\sigma(b), 0)$. Hence, intersecting $\phi(r)$ with the x-axis gives the point $(\sigma(a) + \sigma(b), 0)$. On the other hand, $(a + b, 0)$ is r intersected with the x-axis, so $\phi(a + b, 0) = (\sigma(a + b), 0)$ is $\phi(r)$ intersected with the x-axis (fixed under ϕ). Hence,

$$\sigma(a) + \sigma(b) = \sigma(a+b).$$

By another construction, we can obtain the point $(ba, 0)$ geometrically from the point $(a, 0)$ and $(b, 0)$.

Assume $a \neq 1$ and $b \neq 1$.

(1) Join $(1, 1)$ to $(b, 0)$, call it o.

(2) Draw a line p parallel to o through (a, a).

(3) Intersect p with the x-axis. This is the point $(ba, 0)$ since p has equation $y = -(b - 1)^{-1}x + (b - 1)^{-1}a + a$.

ϕ leaves $(1, 1)$ fixed, so by considering $\phi(o)$ and $\phi(p)$ we can argue as before (Exercise 8.7):

$$\sigma(ba) = \sigma(b)\sigma(a).$$

Hence σ is an automorphism of the division ring R.

Now we return to the projective plane $\mathbf{P}^2(R)$, and show that the effect of ϕ on a point with homogeneous coordinates (x_1, x_2, x_3), is to send it into $(\sigma(x_1,), \sigma(x_2), \sigma(x_3))$ as claimed.

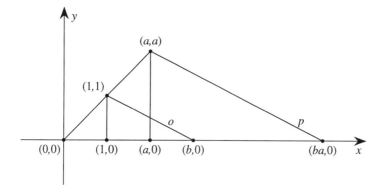

Figure 8.5. $\sigma(b)\sigma(a) = \sigma(ba)$.

Case 1 If $x_3 = 0$, we write this point as the intersection of the line $x_3 = 0$ (which is left fixed by ϕ) and the line joining $(0,0,1)$ with $(x_1, x_2, 1)$. Now the latter point is in **A**, and has affine coordinates (x_1, x_2). Hence ϕ transforms it to $(\sigma(x_1), \sigma(x_2))$, whose homogeneous coordinates are $(\sigma(x_1), \sigma(x_2), 1)$. By intersecting the transformed lines, we find that

$$\phi(x_1, x_2, 0) = (\sigma(x_1), \sigma(x_2), 0).$$

Case 2 If $x_3 \neq 0$, then the point (x_1, x_2, x_3) is in **A**, and has affine coordinates $x = x_1 x_3^{-1}$, $y = x_2 x_3^{-1}$. So

$$\phi(x, y) = (\sigma(x), \sigma(y)) = (\sigma(x_1)\sigma(x_3)^{-1}, \sigma(x_2)\sigma(x_3)^{-1}).$$

(σ takes inverses to inverses by Exercise 7.9.) Therefore $\phi(x, y)$ has homogeneous coordinates $(\sigma(x_1), \sigma(x_2), \sigma(x_3))$ and we are done. □

Since an automorphism ϕ which fixes the standard quadrangle turns out to depend on a division ring automorphism σ, let us rename it and set $\phi = S_\sigma$.

8.16. Proposition *The mapping* $\psi \colon \mathrm{Aut}\, R \to \mathrm{Aut}\, \mathbf{P}^2(R)$ *given by* $\sigma \mapsto S_\sigma$ *is an isomorphism of* $\mathrm{Aut}\, R$ *onto the subgroup* H *of* $\mathrm{Aut}\, \mathbf{P}^2(R)$ *consisting of those automorphisms which leave* P_1, P_2, P_3, P_4 *fixed.*

Proof ψ is onto by the previous proposition. To see it is one-to-one, let σ and $\sigma' \in \mathrm{Aut}\, R$ and apply $S_\sigma, S_{\sigma'}$ to $(x, 1, 0)$. Suppose $(\sigma(x), 1, 0)$ is the same point as $(\sigma'(x), 1, 0)$, then $\sigma(x) = \sigma'(x)$, and $\sigma = \sigma'$. ψ is clearly a homomorphism of groups. □

8.17. Generators Recall that a subset A of a group G is said to generate the subgroup H if H is the smallest subgroup containing A. Then H consists of products of powers of elements in A (Exercise 4.16). In particular, two subgroups H_1, H_2 are said to generate G if their set-theoretic union $H_1 \cup H_2$ generates G.

8.18. Theorem *The two subgroups* $\mathrm{PGL}(2,R)$ *and* H *generate* $\mathrm{Aut}\,\mathbf{P}^2(R)$. *The intersection* K *of* $\mathrm{PGL}(2,R)$ *and* H *is isomorphic to the group* $\mathrm{Inaut}\,R$ *of inner automorphisms of* R.

> **Proof** Given an element $T \in \mathrm{Aut}\,\mathbf{P}^2(R)$, we can find $T_A \in \mathrm{PGL}(2,R)$ such that $T_A(P_i) = T(P_i)$ for $i = 1, 2, 3, 4$ by the fundamental theorem. Then $T_{A^{-1}} \circ T$ fixes P_i so there exists $S_\sigma \in H$ such that $T_{A^{-1}} \circ T = S_\sigma$. Then $T = T_A \circ S_\sigma$ which shows (most strongly) that the subgroups H and $PGL(2,R)$ generate $\mathrm{Aut}\,\mathbf{P}^2(R)$.
>
> An element $T \in \mathrm{PGL}(2,R) \cap H$ fixes the points $P_1, \dots, P_4$ and is induced by an invertible matrix A: i.e. $T = T_A$. By Lemma 8.9, $A = \lambda I$ for some $\lambda \in R$ and, by Lemma 8.10, $T = S_\sigma$ where σ is the inner automorphism given by $\sigma(x) = \lambda x \lambda^{-1}$. $\qquad\square$

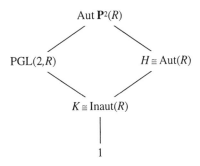

Figure 8.6. The key subgroups in $\mathrm{Aut}\,\mathbf{P}^2(R)$. The corresponding picture for $\mathrm{Aut}\,\mathbf{P}^n(R)$ is the same but with $\mathrm{PGL}(n,R)$, the group of invertible $n \times n$ matrices over R modulo the center, replacing $\mathrm{PGL}(2,R)$.

We are now in a position to determine the automorphism group of the real projective plane.

8.19. Proposition *The identity automorphism is the only automorphism of the field of real numbers.*

> **Proof** Let σ be an automorphism of the real numbers. We proceed in several steps.
> $\sigma(1) = 1$ and $\sigma(a+b) = \sigma(a) + \sigma(b)$. Hence, by induction, $\sigma(n) = n$ for any positive integer n.

$n + (-n) = 0$, so $\sigma(n) + \sigma(-n) = 0$, i.e., $\sigma(-n) = -n$. Hence σ leaves all the integers fixed.

If $b \neq 0$, then $\sigma(a/b) = \sigma(a)/\sigma(b)$. Hence σ leaves all the rational numbers fixed.

If $x \in \mathbf{R}$, then $x > 0$ if and only if there is an $a \neq 0$ such that $x = a^2$. Then $\sigma(x) = \sigma(a)^2$, so $x > 0$ implies $\sigma(x) > 0$: Therefore $x < y$ implies $\sigma(x) < \sigma(y)$, i.e., σ is order-preserving.

Let $x \in \mathbf{R} \smallsetminus \mathbf{Q}$ and suppose $\sigma(x) \neq x$, then either $x < \sigma(x)$ or $x > \sigma(x)$. Suppose $x < \sigma(x)$. By the Archimedean principle there exists a rational number r such that $x < r < \sigma(x)$. Then $r - x > 0$, so that $\sigma(r) - \sigma(x) = r - \sigma(x) > 0$, i.e., $r > \sigma(x)$, a contradiction of the choice of r. Assuming $\sigma(x) < x$ leads to a similar contradiction. Hence $\sigma(x) = x$.

Thus σ is the identity. $\square$

8.20. Theorem $\mathrm{PGL}(2, \mathbf{R}) = \mathrm{Aut}\, \mathbf{P}^2(\mathbf{R})$

Proof By Proposition 8.19, $H = \{1\}$. Then the subgroup $\mathrm{PGL}(2, \mathbf{R})$ is the whole automorphism group, by Theorem 8.18. $\square$

8.21. The fundamental theorem in two dimensions As a consequence of Theorems 8.13 and 8.20 we have the following important theorem for the real projective plane:

8.22. Theorem *Let ABCD and PQRS be complete quadrangles in* $\mathbf{P}^2(\mathbf{R})$. *There exists a unique automorphism of* $\mathbf{P}^2(\mathbf{R})$ *such that* $T(A) = P$, $T(B) = Q$, $T(C) = R$ *and* $T(D) = S$.

You will show in Exercise 8.11 that this theorem is not true in $\mathbf{P}^2(\mathbf{C})$. The problem with $\mathbf{C}$ is that it possesses a nontrivial field automorphism, given by $z \mapsto \bar{z}$. Consequently, the uniqueness part of the theorem does not hold in the general case.

There is a one-dimensional version of this theorem due to von Staudt, the Galician mathematician Ancochea and Hua: see Exercises 7.22 and 10.14.

The fundamental theorems of Chapters 6 and 8 are generalized to n-dimensional real projective spaces in more advanced textbooks (such as [Samuel]) as follows:

Given two ordered sets, X_1 and X_2, of $n + 2$ points of $\mathbf{P}^n(\mathbf{R})$ in general position, i.e., no $n + 1$ points are co-$(n - 1)$-planar, there is a unique automorphism of $\mathbf{P}^n(\mathbf{R})$ sending X_1 into X_2.

8.3   The Algebraic Meaning of Axioms P6 and P7

Now we obtain precise answers for when the axioms P6 and P7 hold in a projective plane $\mathbf{P}^2(R)$.

8.23. Theorem *Fano's axiom P6 holds in* $\mathbf{P}^2(R)$ *if and only if the characteristic of R is* $\neq 2$.

> **Proof** Using an automorphism of $\mathbf{P}^2(R)$, we reduce to the question of whether the diagonal points $(1,1,0)$, $(1,0,1)$, $(0,1,1)$ of the standard quadrangle P_1, P_2, P_3, P_4, are collinear (cf. Proposition 5.2). Since R may not be commutative, we may not use determinants, but must give a hands-on proof.
>
> Suppose they are collinear. Then they all satisfy an equation $c_1x_1 + c_2x_2 + c_3x_3 = 0$, with the c_i not all zero. Hence $c_1 + c_2 = 0$, $c_1 + c_3 = 0$, and $c_2 + c_3 = 0$. Thus $c_1 = -c_2$, $c_1 = -c_3$, $c_2 = -c_3$, so $2c_2 = 0$. So either $c_2 = 0$, in which case $c_3 = 0$, $c_1 = 0$, a contradiction, or $2 = 0$, in which case the characteristic of R is 2.
>
> Now suppose the characteristic of R is 2. Then $(1,1,0)$, $(1,0,1)$, and $(0,1,1)$ satisfy the equation $x_1 + x_2 + x_3 = 0$ of a projective line. $\square$

8.24. Theorem (Hilbert) *The fundamental theorem P7 holds in the projective plane* $\mathbf{P}^2(\mathbf{F})$ *over a division ring* $\mathbf{F}$ *if and only if* $\mathbf{F}$ *is commutative.*

> **Proof** In Chapter 6 we saw that Axiom P7 is equivalent to Pappus' theorem (Theorem 6.6 for FT $\Longrightarrow$ Pappus, Exercises 6.12–6.15 for Pappus $\Longrightarrow$ FT). It will then suffice to prove:
>
> In $\mathbf{P}^2(\mathbf{F})$ the following two conditions are equivalent:
>
> *(1)* For any two lines ℓ, ℓ' and any points $A, B, C \in \ell$, $A', B', C' \in \ell'$, all distinct and different from $X = \ell.\ell'$, the points $P = AB'.A'B$, $Q = AC'.A'C$ and $R = BC'.B'C$ are collinear.
> *(2)* $\mathbf{F}$ is a field.

> Since P, X, C, C' are 4 points in general position (why?), we may find an automorphism sending these into the points $(0,0,1)$, $(1,1,1)$, $(1,0,0)$, and $(0,1,0)$, respectively. We have invoked Theorem 8.13, irrespective of whether $\mathbf{F}$ is a commutative or noncommutative ring, and henceforth assume P, X, C, C' have these coordinates. Now pass to affine coordinates with $x_3 = 0$ the line at infinity. P and X get the affine coordinates $(0,0)$ and $(1,1)$, respectively, while C and C' are ideal points on all horizontal and vertical lines, respectively. Since $XC = \ell$ and $XC' = \ell'$, it follows that ℓ is the line $y = 1$ and ℓ' is the line $x = 1$.

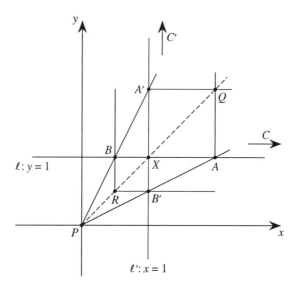

Figure 8.7. $\mathbf{P}^2(\mathbf{F})$ has Pappus' theorem iff $\mathbf{F}$ is a field.

Assign to the points A and B on $y = 1$ the coordinates $(a, 1)$ and $(b, 1)$, respectively. We deduce the coordinates of all the other points through the following simple observation. In $\mathbf{A}^2(\mathbf{F})$, where $\mathbf{F}$ is commutative or not, two points (p, q) and (u, v) are collinear iff $p^{-1}q = u^{-1}v$, since both points lie on a line with equation $y = mx + b$. (If $p = 0$, then $u = 0$, and vice versa.) Since A, B' and P are collinear and B' lies on $x = 1$, B' has coordinates $(1, a^{-1})$. Similarly, the coordinates of A' must be $(1, b^{-1})$.

Now $R = B'C.BC'$ is the intersection of two lines parallel to ℓ and ℓ' in the given affine plane; whence the coordinates of R are deduced from those of B and B' to be (b, a^{-1}). Similarly, $Q = A'C.AC'$ has coordinates (a, b^{-1}). A reapplication of the simple observation above says P, Q and R are collinear iff $a^{-1}b^{-1} = b^{-1}a^{-1}$. This is equivalent to the condition $ab = ba$. This shows that commutativity of multiplication in $\mathbf{F}$ is equivalent to Pappus' theorem. $\qquad\square$

8.4   Independence of Axioms

We are now in a position to show that among the axioms P5, P6, P7, the only implication is that P7 implies P5 (Theorem 6.3). We prove this by giving examples of projective planes which have all relevant combinations of axioms holding or not.

(*1*) The projective plane of seven points π_7 has P5, not P6, P7.

(2) The real projective plane $\mathbf{P}^2(\mathbf{R})$ has P5, P6, P7.

(*3*) The Moulton plane has not P5, P6 (Exercise 6.6), not P7.

(*4*) Let **H** be the division ring of quaternions. Then $\mathbf{P}^2(\mathbf{H})$ has P5, P6, not P7 (because char $\mathbf{H} = 0$ implies P6, and **H** noncommutative implies not P7).

(*5*) Let K be a noncommutative division ring of characteristic 2. Then $\mathbf{P}^2(K)$ has P5, not P6, not P7.

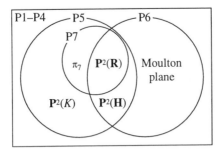

Figure 8.8. Projective planes and Axioms P5–P7.

A couple of things may be seen from the diagram. Axioms P5 and P6 are independent: this means that neither of the implications P5 $\Longrightarrow$ P6 or P6 $\Longrightarrow$ P5 are true. Axioms P6 and P7 are independent as well. There is an example of a projective plane that is neither Fano nor Desarguesian: it is the so-called free projective plane on the π_7 configuration with one extra point (see [Hartshorne, pp. 17–19]).

Exercises

8.1   Prove that Axioms P1–P4 are satisfied in $\mathbf{P}^2(R)$, the projective plane defined by homogeneous coordinates from a division ring R.

8.2   (*a*) Let R be a division ring. Show that the Cartesian product R^2, with lines of form $\{(x,y) \mid y = mx+b\}$ and $\{(x,y) \mid x = a\}$ where $a, m, b \in R$, is an affine plane, denoted by $\mathbf{A}^2(R)$.

(*b*) Complete $\mathbf{A}^2(R)$ to obtain a projective plane S.

(*c*) Adapt Proposition 2.9 and its proof to show that $\mathbf{P}^2(R)$ is the projective plane S.

Hint: Why does a collinearity-preserving one-to-one correspondence (a *collineation*) force one geometry to be a projective plane if the other is so?

8.3   In the real projective plane, we know that there is an automorphism which will send any four points, no three collinear, into any four points, no three collinear. Find the coefficients a_{ij} of an automorphism with equations

$$x_i' = a_{i1}x_1 + a_{i2}x_2 + a_{i3}x_3 \qquad (i = 1, 2, 3)$$

which sends the points $A = (0,0,1)$, $B = (0,1,0)$, $C = (1,0,0)$, $D = (1,1,1)$ into $A' = (1,0,0)$, $B' = (0,1,1)$, $C' = (0,0,1)$, $D' = (1,2,3)$, respectively.

8.4   Let R be a division ring and $A \in M_n(R)$. Let x denote the $n \times n$ matrix

$$\begin{pmatrix} x_1 & 0 & \cdots & 0 \\ x_2 & 0 & \cdots & 0 \\ \vdots & & & \\ x_n & 0 & \cdots & 0 \end{pmatrix}.$$

Suppose that $Ax = 0$ implies $x = 0$ where 0 denotes the zero $n \times n$ matrix. Show that A is invertible.

8.5   Let σ be an automorphism of a division ring R. Check that the mapping $S_\sigma : \mathbf{P}^2(R) \to \mathbf{P}^2(R)$ defined by $S_\sigma : (x_1, x_2, x_3) \mapsto (\sigma(x_1), \sigma(x_2), \sigma(x_3))$ is an automorphism of projective planes.

8.6   Show that Theorem 8.18 implies that every automorphism in $\mathrm{Aut}\,\mathbf{P}^2(R)$ is a semi-linear transformation.

8.7   Provide the details in the proof that $\sigma(ba) = \sigma(b)\sigma(a)$ in Proposition 8.15.

8.8   Complete the details of the proof that $\mathbf{P}^2(F)$ satisfies Pappus' theorem in Hilbert's theorem.

8.9   Prove that $\mathbf{P}^3(R)$, as defined in Theorem 8.5, is a projective 3-space, i.e. satisfies S1–S6.

8.10   If F is a field, re-do Exercise 5.2 in the projective plane $\mathbf{P}^2(F)$. You will now have defined cross ratio for Pappian planes.

8.11   In $\mathbf{P}^2(\mathbf{C})$ consider the standard quadrangle $P_1 P_2 P_3 P_4$. Find two automorphisms of the projective plane $\mathbf{P}^2(\mathbf{C})$ fixing the standard quadrangle (sending $P_i \mapsto P_i$, $i = 1, 2, 3, 4$). Compare with Theorem 8.22.

8.12   As usual, let $\mathbf{Z}_p$ denote the field of integers modulo a prime p.
   (a) Show that the projective plane $\mathbf{P}^2(\mathbf{Z}_p)$ has $p^2 + p + 1$ points.
   (b) Show that a line in $\mathbf{P}^2(\mathbf{Z}_p)$ has $p + 1$ points.
   (c) Compute the order of $\mathrm{Aut}\,\mathbf{P}^2(\mathbf{Z}_p)$.

8.13   Project: Let n be an integer larger than 2. Carry out the analysis of the automorphism group of the n-dimensional projective space over a division ring R suggested in Remark 8.6. You will only need to rewrite the lemmas and propositions of Section 8.2 using $(n+1) \times (n+1)$ matrices over R to arrive at the result in Figure 8.6.
   What is the order of $\mathrm{Aut}\,\mathbf{P}^n(\mathbf{Z}_p)$?

8.14*   Let ℓ, A, B, C and ℓ', A', B', C' be two sets, each consisting of a line and three non-collinear points in a Desarguesian plane π. Prove that there is an automorphism $\alpha : \pi \to \pi$ such that $\alpha(\ell) = \ell'$, $\alpha(A) = A'$, $\alpha(B) = B'$ and $\alpha(C) = C'$.

8.15   By constructing examples, show that no three of the axioms P1, P2, P3, P4 implies the fourth. (Cf. Exercise 2.1.)

Chapter 9

Introduction of Coordinates in a Projective Plane

In this chapter we ask the question, when is a projective plane π isomorphic to a projective plane of the form $\mathbf{P}^2(R)$, for some division ring R? Stated in other words, given π can we find a division ring R, and assign homogeneous coordinates (x_1, x_2, x_3), $x_i \in R$, to points of π, such that the lines are given by linear equations?

A necessary condition for this to be possible is that π should satisfy Desargues' axiom, P5, since we have seen that $\mathbf{P}^2(R)$ always satisfies P5 (Theorem 8.5). And in fact we will see that Desargues' axiom is a sufficient condition that π is coordinatizable on the basis of a division ring.

We will begin with a simpler problem, namely the introduction of coordinates in an affine plane $\mathbf{A}$. One approach to this problem[1] would be the following: Choose three noncollinear points in $\mathbf{A}$, and call them $(1,0)$, $(0,0)$, $(0,1)$. Let ℓ be the line through $(0,0)$ and $(1,0)$. Now take R to be the set of points on ℓ, and define addition and multiplication in R via the geometrical constructions given in the proof of Proposition 8.15. Then one would have to verify that R is a division ring, i.e. prove that addition is commutative and associative, that multiplication is associative and distributive, etc. The proofs involve some rather messy diagrams. Finally one coordinatizes the plane using these coordinates on ℓ, and proves that lines are given by linear equations.

---

[1]done in [Seidenberg], Chapter 3.

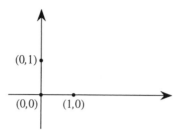

Figure 9.1. One method of coordinatizing.

However, we will use slightly more high-powered techniques, in the hope that our work becomes less onerous. Recall our work with central dilatations and translations in Chapter 1: we will begin by studying the group theory of dilatations and what implications Desargues' theorem has.

9.1   The Major and Minor Desargues' Axioms

Let **A** be an affine plane. Recall that a dilatation of **A** is an automorphism sending each line onto a parallel line. A dilatation different from the identity map on **A** could have either 0 or 1 fixed point (Proposition 1.19): if it has no fixed points, it is called a translation, if it has one fixed point, a central dilatation. However, the identity is considered a translation for practical reasons.

We have proved in Chapter 1 that the set of dilatations, as well as the subset of translations, is closed under composition of maps, contains all inverses, and possesses the identity element, or mapping. The next two propositions are just re-statements of Propositions 1.18 and 1.23 in group-theoretic terms — with one group-theoretic definition intervening.

9.1. Proposition *The set* Dil **A** *of dilatations is a group under composition.* Dil **A** *is a subgroup of* Aut **A**.

9.2. Definition Let G be a group. A subgroup N is said to be *normal* if

$$xNx^{-1} \subseteq N \qquad (\forall x \in G),$$

i.e. $x \in G$ and $y \in N$ implies $xyx^{-1} \in N$.

The kernel of a homomorphism is a normal subgroup (Exercise 9.1). Every subgroup of an abelian group is normal. A normal subgroup N of any group G has equal right and left cosets, and the set of cosets gets a natural group structure (Exercise 9.2)

9.3. Proposition *The set of translations,* Tran **A**, *is a normal subgroup of* Dil **A**.

Now we come to the question of existence of translations and dilatations, and for this we will need Desargues' axiom. In fact, we will find that these two existence problems are equivalent to two affine forms of Desargues' axiom. In this we see yet another example where an axiom about some configuration is equivalent to a geometric property of the plane. Desargues' axiom is equivalent to saying that our geometry has enough automorphisms in a sense which will become clear from the theorems.

> **A4. The minor Desargues' axiom** Let ℓ, m, n be three distinct parallel lines. Let $A, A' \in \ell$, $B, B' \in m$, and $C, C' \in n$. Assume $AB \parallel A'B'$ and $AC \parallel A'C'$. Then $BC \parallel B'C'$.

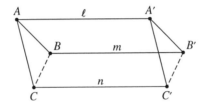

Figure 9.2. The minor Desargues' configuration.

9.4. Example The real affine plane is seen to satisfy A4 by the following vector argument. The vector from B to C, $\boldsymbol{BC} = \boldsymbol{AB} - \boldsymbol{AC} = \boldsymbol{A'B'} - \boldsymbol{A'C'} = \boldsymbol{B'C'}$, whence $BC \parallel B'C'$. (Justify.)

9.5. Example If our affine plane **A** is contained in a projective plane π, then A4 follows from P5 in π. Indeed, ℓ, m, n meet in a point O on the line at infinity ℓ_∞ (cf. Exercise 2.2). Our hypotheses state that $P = AB.A'B' \in \ell_\infty$, and $Q = AC.A'C' \in \ell_\infty$. So P5 says that $R = BC.B'C' \in \ell_\infty$, i.e. $BC \parallel B'C'$ in **A**.

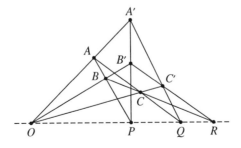

Figure 9.3. *OPQR* is the line at infinity.

9.6. Theorem *Let* **A** *be an affine plane. Then the following statements are equivalent:*

     (*i*) *The axiom A4 holds in* **A**.

     (*ii*) *Given any points* $P, P' \in$ **A**, *there exists a unique translation* τ *such that* $\tau(P) = P'$.

Proof (*ii*) $\implies$ (*i*). We assume the existence of translations, and must deduce A4. Suppose we are given $\ell, m, n, A, A', B, B', C, C'$, as in the statement of A4. Let τ be a translation taking A into A'. Then $\tau(B) = B'$ since $AB \parallel A'B'$ and $AA' \parallel BB'$ (Proposition 1.22). Similarly $\tau(C) = C'$. Hence $BC \parallel B'C'$ since τ is a dilatation.

(*i*) $\implies$ (*ii*). We assume A4. If $P = P'$, then the identity is the only translation taking P to P', so there is nothing to prove.

Now suppose $P \neq P'$. We must construct a translation τ sending P to P'.

Given $Q \notin PP'$, let Q' be the fourth corner of the parallelogram on P, P' and Q (i.e., $Q' = \ell.m$ where $\ell \parallel PP'$, $Q \in \ell$ and $m \parallel PQ$, $P' \in m$). Define $\tau(Q) = Q'$. This defines τ everywhere but on the line PP'.

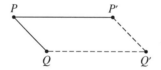

Figure 9.4. Parallelogram construction of Q'.

Given $X \in PP'$, let X' be the fourth corner of the parallelogram on X, Q and Q'. Define $\tau(X) = X'$.

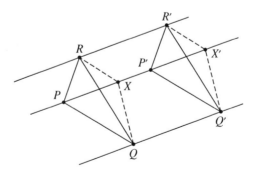

Figure 9.5. τ defined on PP'.

τ is unambiguously defined, for given R and R' such that $RR' \parallel PP'$ and $PR \parallel P'R'$ (and hence $\tau(R) = R'$), it follows from A4 that $RQ \parallel R'Q'$. Since $XQ \parallel X'Q'$, we have $RX \parallel R'X'$ as well by A4. Hence, X' is obtained again after replacing Q, Q' with R, R'.

It remains to show that τ is a translation. Clearly, τ^{-1} is defined by sending P' into P, Q' into Q, etc., by again using the parallelogram construction, whence τ is a bijection.

We must check that collinear points X, Y, Z are mapped into collinear points X', Y', Z' by τ. By A4 applied to the parallel lines PP', XX' and YY', we have $XY \parallel X'Y'$. By A4 applied to the parallel lines PP', YY' and ZZ', we have $YZ \parallel Y'Z'$. But $XY = YZ$, so $X'Y' \parallel Y'Z'$. Hence $X'Y' = Y'Z'$.

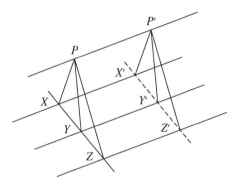

Figure 9.6. τ is an automorphism.

τ is a dilatation, for given R, S, we find from A4 that $RS \parallel R'S'$. (Check the other cases.) It is clear from the parallelogram construction that τ has no fixed point.

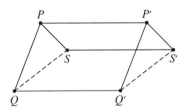

Figure 9.7. τ is a dilatation.

The uniqueness of τ: if τ' is another translation sending P to P', then $\tau'\tau^{-1}$ has a fixed point and is the identity. So $\tau = \tau'$.   □

9.7. Proposition *Assuming A4,* Tran **A** *is an abelian group.*

Proof Let τ, τ' be translations. We must show that $\tau\tau' = \tau'\tau$.

Case 1 Suppose τ and τ' translate in different directions. Let P be a point. Let $\tau(P) = P'$, $\tau'(P) = Q$. We are assuming P, P', and Q are not collinear. Then $\tau(Q) = \tau\tau'(P)$ and $\tau'(P') = \tau'\tau(P)$ are both found as the fourth vertex of the parallelogram on P, P', Q, hence are equal, so $\tau\tau' = \tau'\tau$. (So far we have not used Axiom A4.)

Case 2 Suppose τ and τ' are in the same direction. By Theorem 9.6 there exists a translation σ in a different direction (Axiom A3 ensures that there is another direction). Then

$$\tau\tau' = \tau(\tau'\sigma)\sigma^{-1} = (\tau'\sigma)\tau\sigma^{-1} = \tau'(\sigma\tau)\sigma^{-1}$$

by an application of Case 1, since τ and $\tau'\sigma$ are in different directions. Then, since τ and σ are in different directions, we may transpose the symbols within parentheses, so

$$\tau\tau' = \tau'\tau\sigma\sigma^{-1} = \tau'\tau.$$

$\square$

9.8. Definition We say a group G is the *semi-direct product* of two subgroups H and K, and write $G = H \rtimes K$, if

> **SD1.** H is a normal subgroup of G.
> **SD2.** $H \cap K = \{1\}$.
> **SD3.** H and K together generate G.

9.9. Proposition *If $G = H \rtimes K$, then every element $g \in G$ can be written uniquely as a product $g = hk$, $h \in H$, $k \in K$.*

Proof Note that $hkh_1k_1 = h(kh_1k^{-1})kk_1$ is of the form $h'k'$, $h' \in H$, $k' \in K$, since H is a normal subgroup of G. Any element in G may be written as a product of elements from H and K by Axiom SD3, hence can be put in the form $h'k'$.

Uniqueness follows from the observation that $hk = h_1k_1$ implies $h_1^{-1}h = k_1k^{-1}$, whence $h_1^{-1}h = 1 = k_1k^{-1}$ by SD2, so $h = h_1$ and $k = k_1$. $\square$

9.10. Definition Let O be a point in $\mathbf{A}$, and define $\mathrm{Dil}_O(\mathbf{A})$ to be the subset of $\mathrm{Dil}\,\mathbf{A}$ consisting of those dilatations ϕ such that $\phi(O) = O$. It is trivial to see that $\mathrm{Dil}_O(\mathbf{A})$, the set of central dilatations fixing O, is a group.

9.11. Proposition *Assuming A4, $\mathrm{Dil}\,\mathbf{A}$ is the semi-direct product of $\mathrm{Tran}\,\mathbf{A}$ and $\mathrm{Dil}_O(\mathbf{A})$.*

Proof We agree that $\mathrm{Tran}\,\mathbf{A}$ and $\mathrm{Dil}_O(\mathbf{A})$ are subgroups of the group $\mathrm{Dil}\,\mathbf{A}$. We need to check the three axioms SD1–SD3.

(*1*) We have seen that $\mathrm{Tran}\,\mathbf{A}$ is a normal subgroup of $\mathrm{Dil}\,\mathbf{A}$.

(*2*) If $\tau \in \mathrm{Tran}\,\mathbf{A} \cap \mathrm{Dil}_O(\mathbf{A})$, then τ has the fixed point O and must be the identity.

(*3*) Let $\phi \in \mathrm{Dil}\,\mathbf{A}$. Let Q denote $\phi(O)$. By Theorem 9.6 there is a translation τ such that $\tau(O) = Q$. Then $\tau^{-1}\phi \in \mathrm{Dil}_O(\mathbf{A})$, so $\phi = \tau(\tau^{-1}\phi)$ shows that $\mathrm{Tran}\,\mathbf{A}$ and $\mathrm{Dil}_O(\mathbf{A})$ generate $\mathrm{Dil}\,\mathbf{A}$.   □

A5. The major Desargues' axiom Let O, A, B, C, A', B', C' be distinct points in the affine plane **A**, and assume that O, A, A' are collinear, O, B, B' are collinear, O, C, C' are collinear, $AB \parallel A'B'$, and $AC \parallel A'C'$. Then $BC \parallel B'C'$.

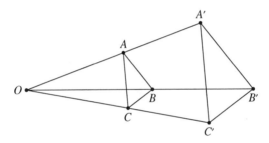

Figure 9.8. Major Desargues' axiom.

Note that this statement follows from P5, if **A** is embedded in a projective plane π.

9.12. Theorem *The following two statements are equivalent in the affine plane* **A**.

(*i*) *The axiom A5 holds in* **A**.

(*ii*) *Given any three points O, P, P', with $P \neq O$, $P' \neq O$, and O, P, P' collinear, there exists a unique dilatation σ of **A**, such that $\sigma(O) = O$ and $\sigma(P) = P'$.*

Proof (*ii*) $\implies$ (*i*). Let O, A, B, C, A', B', C' be given satisfying the hypothesis of A5. Let σ be a dilatation which leaves O fixed and sends A into A'. Then $\sigma(B) = B'$ since $\sigma(OB) \parallel OB$ implies $\sigma(OB) = OB$, and $\sigma(AB) \parallel AB$. Similarly, $\sigma(C) = C'$. Since σ is a dilatation, $BC \parallel B'C'$.

(i) $\implies$ (ii). Given O, P, P', as above, define a transformation σ, for points Q not on the line ℓ containing O, P, P' as follows: $\sigma(Q) = Q'$, where Q' is the intersection of the line OQ with the line through P', parallel to PQ.

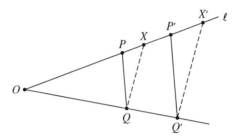

Figure 9.9. Trapezoid construction of $Q' = \sigma(Q)$ and $X' = \sigma(X)$.

Given $X \in PP'$, let $\sigma(X) = X'$ be the intersection of ℓ with the line parallel to XQ through Q'. The rest of the proof is similar to the proof of Theorem 9.6, just with the trapezoid replacing the parallelogram (Exercise 9.4). □

We make a little diversion into the relationship of Axioms A4 and A5 with the next proposition.

9.13. Proposition *A5 implies A4.*

Proof Indeed, let us assume we have an affine plane **A** satisfying A5. Let P, P' be two points. We will construct a translation sending P into P', which by Theorem 9.6 shows that A4 holds, since P, P' are arbitrary. If $P = P'$, we can take the identity, so assume from the start that $P \neq P'$.

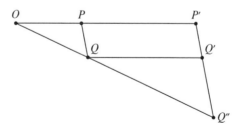

Figure 9.10. A5 $\implies$ A4.

Let Q be a point not on PP', and let Q' be the fourth vertex of the parallelogram on P, P', Q. Let O be a point on PP', not equal to P or P'. Let σ_1 be a dilatation which leaves O fixed, and sends P to P' (which exists by Theorem

9.12). Let $\sigma_1(Q) = Q''$. Then P', Q', Q'' are collinear. Again, by Theorem 9.12 there exists a dilatation σ_2 leaving P' fixed, and sending Q'' to Q'.

Now consider $\tau = \sigma_2\sigma_1$. Being a product of dilatations, it is itself a dilatation. One sees easily that $\tau(P) = P'$ and $\tau(Q) = Q'$. Now any fixed point of τ must lie on PP' and on QQ' (because if X is a fixed point, then $XP \parallel XP'$, whence X, P, P' collinear, and similarly for Q). But $PP' \parallel QQ'$, so τ has no fixed points. Hence τ is a translation sending P into P'.   □

9.14. Remark A projective plane is called a *translation plane* (or *alternative plane*) if, upon removing any line, the resulting affine plane satisfies Desargues' minor axiom. (Compare with page 150.) Translation planes and their automorphism groups have been an active area of research in recent memory.

9.2   Division Ring Number Lines

Now we come to the construction of coordinates in an affine plane **A** satisfying Axioms A4 and A5. Our program is to construct the following objects:

> *(1)* a division ring R;
> *(2)* coordinates for the points of **A** so that **A** is in one-to-one correspondence with the set of ordered pairs of elements of R;
> *(3)* equations for an arbitrary translation and an arbitrary dilatation of **A** in terms of coordinates;
> *(4)* linear equations for the lines in **A**.

This will prove that **A** is isomorphic to the affine plane $\mathbf{A}^2(R)$, with *(3)* a convenient bonus.

In the course of these constructions, there will be about a hundred details to verify, so we will not attempt to do them all, but will give indications, and leave the trivial verifications to the reader.

Fix a line ℓ in **A**, and fix two points on ℓ, call them 0 and 1. Now let R be the set of points of ℓ.

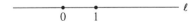

Figure 9.11. The number line.

Let $a \in R$ (i.e., a is a point of ℓ). By A4 and Theorem 9.6, there is a unique translation τ_a that takes 0 into a. Similarly, using A5 and assuming $a \neq 0$, let σ_a be the unique dilatation of **A** which leaves 0 fixed, and sends 1 into a.

Now we define addition and multiplication in R as follows. If $a, b \in R$, define

$$a + b = \tau_a\tau_b(0) = \tau_a(b). \tag{9.1}$$

Since the translations form an abelian group, we see immediately that addition is associative and commutative:

$$(a+b)+c = a+(b+c),$$
$$a+b = b+a.$$

Since $\tau_0 = \mathrm{id}$ we see that 0 is the identity element. Let $-a = \tau_a^{-1}(0)$: this is clearly the inverse of a. Thus R is an abelian group under addition.

Translations are equal if they agree on one point, so

$$\tau_{a+b} = \tau_a \tau_b \qquad \text{for all } a,b \in R. \tag{9.2}$$

Now we define multiplication as follows: first, 0 times anything is 0. Second, if $a,b \in R$, $b \neq 0$, we define

$$ab = \sigma_b(a) = \sigma_b\sigma_a(1). \tag{9.3}$$

Now, since the dilatations form a group, we see immediately that

$$(ab)c = a(bc),$$

that $a \cdot 1 = 1 \cdot a = a$ for all a, and that $\sigma_a^{-1}(1) = a^{-1}$ is a multiplicative inverse. Therefore the nonzero elements of R form a group under multiplication. Furthermore, we have the formulae (for $b \neq 0$)

$$\tau_{ab} = \sigma_b \tau_a \sigma_b^{-1} \tag{9.4}$$

$$\sigma_{ab} = \sigma_b \sigma_a, \tag{9.5}$$

the top equation follows by checking it on 0; the bottom on 0 and 1.

It remains to establish the distributive laws in R. The left distributive law is much harder than the right, perhaps because our definition of multiplication is asymmetric. First consider $(a+b)c$. If $c = 0$, $(a+b)c = 0 = ac + bc$. If $c \neq 0$, we use formulas 9.4 and 9.2, and find $\tau_{(a+b)c} = \sigma_c \tau_{a+b} \sigma_c^{-1} = \sigma_c \tau_a \tau_b \sigma_c^{-1} = \sigma_c \tau_a \sigma_c^{-1} \sigma_c \tau_b \sigma_c^{-1} = \tau_{ac} \tau_{bc} = \tau_{ac+bc}$. Now applying both ends of this equality to the point 0, we have

$$(a+b)c = ac + bc.$$

Before proving the left distributive law, we must establish a lemma. For any line m in $\mathbf{A}$, denote the group of translations in the direction of m by $\mathrm{Tran}_m(\mathbf{A})$, i.e. those translations $\tau \in \mathrm{Tran}\,\mathbf{A}$ such that either $\tau = \mathrm{id}$ or $PP' \parallel m$ for all P (where $\tau(P) = P'$).

Let m,n be lines in $\mathbf{A}$ (which may be the same). Let $\tau' \in \mathrm{Tran}_m(\mathbf{A})$ and $\tau'' \in \mathrm{Tran}_n(\mathbf{A})$ be fixed translations, different from the identity, and let o be a fixed point of **A**. We define a mapping $\phi \colon \mathrm{Tran}_m(\mathbf{A}) \to \mathrm{Tran}_n(\mathbf{A})$ as follows. For each $\tau \in$ $\mathrm{Tran}_m(\mathbf{A})$, $\tau \neq \mathrm{id}$, there exists by Theorem 9.12 a unique central dilatation σ leaving o fixed such that $\sigma(\tau'(o)) = \tau(o)$. Then $\tau = \sigma\tau'\sigma^{-1}$ since both τ and $\sigma\tau'\sigma^{-1}$ are translations that agree on o. Now using σ define ϕ by

$$\phi(\tau) = \sigma\tau''\sigma^{-1}. \tag{9.6}$$

Check that $\sigma\tau''\sigma^{-1}$ is a translation in the direction of n (Exercise 9.6).

9.15. Lemma $\phi \colon \mathrm{Tran}_m(\mathbf{A}) \to \mathrm{Tran}_n(\mathbf{A})$ *is a homomorphism of groups.*

Proof We need to show that $\phi(\tau_1\tau_2) = \phi(\tau_1)\phi(\tau_2)$ for all $\tau_1, \tau_2 \in \mathrm{Tran}_m(\mathbf{A})$. There is a subtlety in this since σ depends on the argument. We distinguish two cases.

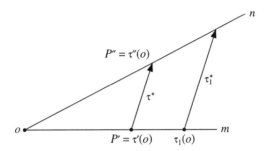

Figure 9.12. σ_1 stretches P' into $\tau_1(o)$.

Case 1 Suppose $m \nmid n$. Replacing m and n by lines parallel to them, if necessary, we may assume that m and n pass through o. Let $\tau'(o) = P'$ and $\tau''(o) = P''$. Let τ^* be the unique translation which takes P' into P''. Then

$$\tau'' = \tau'\tau^*.$$

If $\tau_1, \tau_2 \in \mathrm{Tran}_m(\mathbf{A})$, let σ_1, σ_2 be the corresponding central dilatations, i.e. $\tau_i = \sigma_i\tau'\sigma_i^{-1}$ $(i = 1, 2)$. Then

$$\phi(\tau_1) = \sigma_1\tau''\sigma_1^{-1} = \sigma_1\tau'\tau^*\sigma_1^{-1} = \sigma_1\tau'\sigma_1^{-1}\sigma_1\tau^*\sigma_1^{-1} = \tau_1 \circ \sigma_1\tau^*\sigma_1^{-1} = \tau_1\tau_1^*,$$

where we define $\tau_1^* = \sigma_1\tau^*\sigma_1^{-1}$. Similarly, $\phi(\tau_2) = \tau_2\tau_2^*$, where $\tau_2^* = \sigma_2\tau^*\sigma_2^{-1}$. If σ_3 is the dilatation with center o such that $\tau_1\tau_2 = \sigma_3\tau'\sigma_3^{-1}$, we get by a similar calculation that $\phi(\tau_1\tau_2) = \tau_1\tau_2 \circ \tau_3^*$, where $\tau_3^* = \sigma_3\tau^*\sigma_3^{-1}$.

So we have

$$\phi(\tau_1\tau_2) = \tau_1\tau_2 \circ \tau_3^* \quad \text{and} \quad \phi(\tau_1)\phi(\tau_2) = \tau_1\tau_2 \circ \tau_1^*\tau_2^*.$$

Let $Q = \phi(\tau_1\tau_2)(o)$ and $R = \phi(\tau_1)\phi(\tau_2)(o)$. Now $\phi(\tau_1\tau_2)$ and $\phi(\tau_1)\phi(\tau_2)$ are both translations in the m direction. Then o, Q and R are collinear. But τ_3^* and $\tau_1^*\tau_2^*$ are both translations in the τ^* direction, so $\tau_1\tau_2(o)$, Q and R are collinear. Hence $Q = R$. It follows that $\phi(\tau_1\tau_2) = \phi(\tau_1)\phi(\tau_2)$.

Case 2 Suppose $m \parallel n$. Let $\tau', \tau'' \in \mathrm{Tran}_m(\mathbf{A})$. Take another line p not parallel to m, and take $\tau''' \in \mathrm{Tran}_p(\mathbf{A})$. Define $\psi_1 : \mathrm{Tran}_m(\mathbf{A}) \to \mathrm{Tran}_p(\mathbf{A})$ using τ' and τ'', and define $\psi_2 : \mathrm{Tran}_p(\mathbf{A}) \to \mathrm{Tran}_m(\mathbf{A})$ using τ''' and τ''.

Then $\phi = \psi_2\psi_1$ (Exercise 9.7). But ψ_1, ψ_2 are homomorphisms by Case 1. Hence, ϕ is a homomorphism.   □

Now we can prove the left distributivity law as follows. Consider $\lambda(a + b)$. In the lemma, take $m = n = \ell$, $o = 0$, $\tau' = \tau_1$, $\tau'' = \tau_\lambda$. Then ϕ is the map of $\mathrm{Tran}_\ell(\mathbf{A}) \to \mathrm{Tran}_\ell(\mathbf{A})$ which sends τ_a into $\tau_{\lambda a}$, for any a. Indeed, by Equation 9.4, $\tau_a = \sigma_a\tau_1\sigma_a^{-1}$, so $\phi(\tau_a) = \sigma_a\tau_\lambda\sigma_a^{-1} = \tau_{\lambda a}$. By the lemma, $\phi(\tau_a\tau_b) = \phi(\tau_a)\phi(\tau_b)$ for all $a, b \in R$, whence $\phi(\tau_{a+b}) = \phi(\tau_a)\phi(\tau_b)$ by Equation 9.2. Hence $\tau_{\lambda(a+b)} = \tau_{\lambda a}\tau_{\lambda b} = \tau_{\lambda a + \lambda b}$. Evaluating both translations at 0, we get by Equation 9.2

$$\lambda(a + b) = \lambda a + \lambda b.$$

This completes the proof of

9.16. Theorem *Let $\mathbf{A}$ be an affine plane satisfying Axioms A4 and A5. Let ℓ be a line of $\mathbf{A}$, let $0, 1$ be two points of ℓ, let R be the set of points of ℓ, and define $+$ and $\cdot$ in R as given in Equations 9.1 and 9.3 above. Then R is a division ring.*

9.3   Introducing Coordinates in A

We are now ready to introduce coordinates in $\mathbf{A}$. We have already fixed a line ℓ in $\mathbf{A}$, and two points $0, 1$ on ℓ. On the basis of these choices we defined our division ring R. Choose another line, m, passing through 0, and fix a point $1'$ on m.

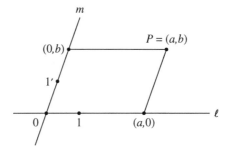

Figure 9.13. Assigning coordinates to points in $\mathbf{A}$.

For each point $P \in \ell$, if P corresponds to the element $a \in R$, we give P the affine coordinates $(a, 0)$. Thus 0 and 1 have coordinates $(0,0)$ and $(1,0)$, respectively.

If $P \in m$, $P \neq 0$, then there is a unique dilatation σ leaving 0 fixed and sending $1'$ into P. σ is of the form σ_a where $a = \sigma(1) \in R$. So we give P coordinates $(0, a)$.

Finally, if P is a point not on ℓ or m, we draw lines through P, parallel to ℓ and m, to intersect m in $(0, b)$ and ℓ in $(a, 0)$. Then we give P the coordinates (a, b).

One sees easily that in this way $\mathbf{A}$ is put into one-to-one correspondence with the set of ordered pairs of elements of R. We have yet to see that lines are given by linear equations — this will come after we find the equations of translations and dilatations.

We investigate the equations of translations and dilatations. First some notation. For any $a \in R$, denote by τ'_a the translation which takes 0 into $(0, a)$. Thus τ'_1 is the translation which takes 0 into $1'$, and for any $a \in R$, $a \neq 0$,

$$\tau'_a = \sigma_a \tau'_1 \sigma_a^{-1}. \tag{9.7}$$

This follows from the definition of the point $(0, a)$. Furthermore, it follows from Equation 9.7 and Lemma 9.15 that the mapping $\phi: \mathrm{Tran}_\ell(\mathbf{A}) \to \mathrm{Tran}_m(\mathbf{A})$ defined by $\tau_a \mapsto \tau'_a$ is a homomorphism, and hence we have the formulas, for any $a, b \in R$,

$$\tau'_{a+b} = \tau'_a \tau'_b \tag{9.8}$$

$$\tau'_{ab} = \sigma_b \tau'_a \sigma_b^{-1}, \tag{9.9}$$

the bottom formula (Equation 9.9) coming from applying σ_b on the left, and σ_b^{-1} on the right of Equation 9.7 and recalling Equation 9.5.

9.17. Proposition *Let τ be a translation of $\mathbf{A}$, and suppose that $\tau(0) = (a, b)$. Then τ takes an arbitrary point $Q = (x, y)$ into $Q' = (x', y')$ where*

$$\begin{cases} x' = x + a \\ y' = y + b. \end{cases}$$

Proof Indeed, let τ_Q be the translation taking 0 into Q. Then $\tau_Q = \tau_x \tau'_y$. Also $\tau = \tau_a \tau'_b$. So $\tau(Q) = \tau \tau_Q(0) = \tau_a \tau'_b \tau_x \tau'_y(0) = \tau_a \tau_x \tau'_b \tau'_y(0) = \tau_{a+x} \tau'_{b+y}(0) = (x + a, y + b)$. $\qquad\square$

9.18. Proposition *Let σ be any dilatation of $\mathbf{A}$ leaving 0 fixed. Then $\sigma = \sigma_a$ for some $a \in R$, and σ takes the point $Q = (x, y)$ into $Q' = (x', y')$, where*

$$\begin{cases} x' = xa \\ y' = ya. \end{cases}$$

Proof Again write $\tau_Q = \tau_x \tau_y'$. Then, using Equations 9.2 and 9.9,

$$\sigma(Q) = \sigma_a \tau_x \tau_y'(0) = \sigma_a \tau_x \tau_y' \sigma_a^{-1}(0)$$

$$= \sigma_a \tau_x \sigma_a^{-1} \cdot \sigma_a \tau_y' \sigma_a^{-1}(0) = \tau_{xa} \cdot \tau_{ya}'(0) = (xa, ya). \qquad \square$$

9.19. Theorem *Let* **A** *be an affine plane satisfying A4 and A5. Fix two nonparallel lines ℓ, m in* **A**, *and fix points $1 \in \ell$ and $1' \in m$, different from $0 = \ell.m$. Then assigning coordinates as above, the lines in* **A** *are all linear equations of the form $y = mx + b$, $m, b \in R$, or $x = a$, $a \in R$. Thus* **A** *is isomorphic to the affine plane* $\mathbf{A}^2(R)$.

Proof By construction of the coordinates, a line parallel to ℓ will have an equation of the form $y = b$, and a line parallel to m wil have an equation of the form $x = a$.

Now let r be any line through 0, different from ℓ and m. Then r must intersect the line $x = 1$, say in the point $Q = (1, a)$ for some $a \in R$.

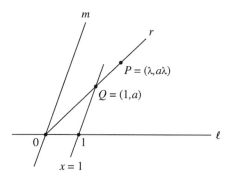

Figure 9.14. Assigning a linear equation to r.

Now if P is any other point on r, different from 0, there is a unique dilatation σ_λ leaving 0 fixed and sending Q into P. Hence P will have coordinates $x = 1 \cdot \lambda$, $y = a \cdot \lambda$. Substituting x for λ, we find the equation of r is $y = ax$.

Finally, let s be a line not passing through 0, and not parallel to ℓ or m. Let r be the line parallel to s, passing through 0. Let s intersect m in $(0, b)$. Then it is clear that the points of s are obtained by applying the translation τ_b' to the points of r. So if $(\lambda, a\lambda)$ is a point of r (for $x = \lambda$), the corresponding point of s will be $x = \lambda + 0 = \lambda$, $y = a\lambda + b$, by Proposition 9.17. So the equation of s is $y = ax + b$.

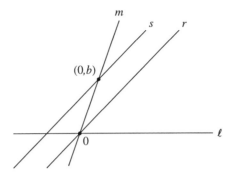

Figure 9.15. Assigning an equation to s.

$\square$

9.20. Remark If σ is an arbitrary dilatation of $\mathbf{A}$, then σ can be written as $\tau\sigma'$, where τ is a translation, and σ' is a dilatation leaving 0 fixed (cf. Proposition 9.11). So if τ has equations $x' = x + c$, $y' = y + d$, and σ' has equations $x' = xa$, $y' = ya$, we find that σ has equations

$$\begin{cases} x' = xa + c \\ y' = ya + d \end{cases}$$

9.21. Theorem *Let π be a projective plane satisfying P1–P5. Then there is a division ring R such that π is isomorphic to $\mathbf{P}^2(R)$, the projective plane over R.*

Proof Let ℓ_0 be any line in π, and consider the affine plane $\mathbf{A} = \pi \smallsetminus \ell_0$. Then $\mathbf{A}$ satisfies A4 and A5, hence $\mathbf{A} \cong \mathbf{A}^2(R)$, by the previous theorem. But π is the projective plane obtained by completing the affine plane $\mathbf{A}$, and $\mathbf{P}^2(R)$ is the projective plane that completes the affine plane $\mathbf{A}^2(R)$ (Exercise 8.2), so the isomorphism above extends to show $\pi \cong \mathbf{P}^2(R)$. $\square$

9.22. Remark This is a good point at which to clear up a question left hanging from the early chapters, about the correspondence between affine planes and projective planes. We saw that an affine plane $\mathbf{A}$ could be completed to a projective plane $S(\mathbf{A})$ by adding ideal points and an ideal line. Conversely, if π is a projective plane, and ℓ_0 a line in π then $\pi \smallsetminus \ell_0$ is an affine plane, by Exercise 2.2.

What happens if we perform first one process and then the other? Do we get back to where we started? There are two cases to consider.

(*1*) If π is a projective plane, ℓ a line in π, and $\pi \smallsetminus \ell$ the corresponding affine plane, then one can see easily (Exercise 2.2) that $S(\pi \smallsetminus \ell)$ is isomorphic to π in a natural way.

(2) Let $\mathbf{A}$ be an affine plane, and let $S(\mathbf{A}) = \mathbf{A} \cup \ell_\infty$ be the corresponding projective plane. Then clearly $S(\mathbf{A}) \smallsetminus \ell_\infty \cong \mathbf{A}$. But what if ℓ_1 is a line in $S(\mathbf{A})$ different from ℓ_∞? Then in general one cannot expect $S(\mathbf{A}) \smallsetminus \ell_1$ isomorphic to $\mathbf{A}$.

However, if we assume that $\mathbf{A}$ satisfies A4 and A5, then $S(\mathbf{A}) \smallsetminus \ell_1 \cong \mathbf{A}$. Indeed, $S(\mathbf{A}) \cong \mathbf{P}^2(R)$, for some division ring R, and we can always find an automorphism $\phi \in \operatorname{Aut} \mathbf{P}^2(R)$, taking ℓ to ℓ_∞ (see Exercise 8.14). Then ϕ gives an isomorphism of $S(\mathbf{A}) \smallsetminus \ell_1$ and $\mathbf{A}$.

9.23. Remark We may now draw a conclusion from our efforts in this chapter. In Chapters 2, 5 and 6, we gave seven axioms, P1–P7, for a planar projective geometry. By the end of Chapter 8, we understood that the projective planes over fields, defined earlier in the same chapter, provide models for our system of axioms, so long as the characteristic of the field is $\neq 2$.

Are these all the models? The answer is now seen to be in the affirmative, since a projective plane with P5 has been shown to be necessarily $\mathbf{P}^2(R)$ for some division ring R. If P6 and P7 are to be met, R must be a field of characteristic $\neq 2$. We may say that the axioms P1–P4, P6 and P7 *characterize* projective planes over fields of characteristic $\neq 2$, since P7 $\implies$ P5 (Theorem 6.3).

In proving that a certain axiom system is consistent (i.e., is free from logical inconsistencies), one shows that the axiom system has a certain acceptable model, which meets all the conditions of the axioms. Any inconsistencies in the axiom system would then be an inconsistency in the model. The converse is true if the model turns out to be the *only* model for the axiom system. Hence, we conclude that our synthetic theory of projective planes satisfying P1–P7 is as consistent as the analytic theory of projective planes over fields of characteristic $\neq 2$.

Exercises

9.1   Prove that the kernel of a homomorphism of groups is a normal subgroup.

9.2   In this exercise you will assist in the definition of *factor group* G/N obtained from a group G and normal subgroup N. As a set, let $G/N = \{gN \mid g \in G\}$, the set of left cosets of N in G. (Beware that the same coset may go under different names: $gN = g'N$ so long as $g^{-1}g' \in N$.)

    (*a*) Show that for each g in G the right coset of g equals the left coset of g: i.e. $gN = Ng$.

    (*b*) Show that the following formula defines a group operation on G/N:

$$(gN)(g'N) = gg'N.$$

Where must you use normality of N?

    (*c*) If G is a semidirect product of N with another subgroup K, i.e. $G = N \rtimes K$, find an isomorphism showing $G/N \cong K$.

9.3   A group is said to be *simple* if it has no proper normal subgroups. [2]
(*a*) If G is simple and $f\colon G \to K$ is a homomorphism, show that f is either injective or constant.
(*b*) Show that the additive groups $\mathbf{Z}_p$, p a prime, are simple.
(*c**) Project: Read [Herstein, chapter 2] and [Chapman]. Show that the 168-element group $\operatorname{Aut}\mathbf{P}^2(\mathbf{Z}_2)$ is simple. Do the same for $\operatorname{Aut}\mathbf{P}^2(\mathbf{Z}_3)$ and $\operatorname{Aut}\mathbf{P}^2(\mathbf{Z}_5)$.

9.4   Give a complete proof of Theorem 9.12.

9.5   Give a rigorous proof of Case 2 in the proof of Proposition 9.7.

9.6   Prove that $\sigma\tau''\sigma^{-1}$ in Equation 9.6 is a translation in the direction of n.

9.7   Prove that $\phi = \psi_2\psi_1$ in Case 2 of the proof of Lemma 9.15.

9.8   Let π be a Desarguesian plane. Theorem 9.21 tells us that we may coordinatize points in π by selecting a line ℓ_0, and coordinatizing the affine plane $\mathbf{A} = \pi \smallsetminus \ell_0$. $\mathbf{A}$ may be coordinatized by selecting nonparallel lines ℓ_1 and ℓ_2 in $\mathbf{A}$, $O = \ell_1.\ell_2$, $1 \in \ell_1$, $1' \in \ell_2$, $R = \{\text{points on } \ell_1\}$, $+$ and $\cdot$ defined in Section 9.2, and proceeding as in Section 9.3.
(*a*) Show how to assign coordinates (x_1, x_2, x_3) to each point of π.
(*b*) Show how to assign a linear equation $\sum_{i=1}^{3} c_i x_i = 0$ to each line in π.

9.9   Is the homomorphism $\phi\colon \operatorname{Tran}_m(\mathbf{A}) \to \operatorname{Tran}_n(\mathbf{A})$ actually an isomorphism of groups?

9.10   In Section 9.2 we let R be the set of points on a line ℓ in $\mathbf{A}$, where two points are designated 0 and 1. Addition and multiplication are defined by means of translation along ℓ and central dilatation fixing O (cf. Equations 9.1 and 9.3). Suppose m is a line in $\mathbf{A}$ intersecting ℓ in 0. Take R' to be the set of points with 0 and a point $1'$ to be chosen in $R' \smallsetminus \{0\}$. Define addition by translations along m and Equation 9.1, multiplication by central dilatations fixing 0 and Equation 9.3. Then R and R' are division rings.
(*a*) Show that they are isomorphic rings: $R \cong R'$.
(*b*) If m were parallel to ℓ, $0'$ and $1'$ arbitrary points in m, show that $R \cong R'$.
(*c*) Show that $R \cong R'$ with no restriction on m.

---

[2]Simple groups are the basic building blocks of group theory. The Jordan-Hölder theorem (in [Jacobson]) states that any finite group determines a unique list of simple groups, whose cardinality is called the length of the group. Simple groups have length 1. The guiding principle is that much information about a group G is retrieved from looking at the smaller groups N, where N is a normal subgroup of G, and the factor group G/N. Equivalently, we look at the homomorphisms with domain G for information about G. Simple groups are then the indivisible blocks in this theory by Exercises 9.1–9.3.

Projective geometry and its generalizations have provided many of the examples of finite simple groups. In the 1970's, researchers felt that all examples of finite simple groups had been found. They set about proving this, and achieved in the 1980's the complete classification of finite simple groups. The present decade has seen several old conjectures about finite groups settled on the basis of this classification.

Chapter 10

Möbius Transformations and Cross Ratio

10.1 Assessment

Let us pause for a moment to see what we have done and where we are going. We have been studying the subject of projective geometry from two points of view, the synthetic and the analytic.

The synthetic approach to planar projective geometry starts with points and lines satisfying Axioms P1–P4. We make definitions like automorphism and complete quadrangle, proceeding in logical steps and proving theorems. Eventually we add Axioms P5, P6 and P7 as we need them. For example, we add P6 when we need harmonic points, and P5 when we need to show the well-definedness of the harmonic conjugate. We added P7 in order to make the group $\mathrm{PJ}(\ell)$ of conjective projectivities precisely 3-transitive, and prove Pappus' theorem.

The analytic approach to projective geometry starts from an algebraic object like the reals, complex numbers or any division ring R or field $\mathbf{F}$. We defined $\mathbf{P}^2(\mathbf{F})$ as nonzero ordered triples of $\mathbf{F}$-elements with equivalence relation $(x_1, x_2, x_3) \sim (x_1, x_2, x_3)\lambda$, and lines as linear equations. We defined cross ratio,[1] a certain group of automorphisms, viz. $\mathrm{PGL}(2, \mathbf{F})$, using 3×3 matrices, another group using field automorphisms of $\mathbf{F}$, and proved a fundamental theorem telling that these two subgroups together generate $\mathrm{Aut}\,\mathbf{P}^2(\mathbf{F})$. In addition, two ordered sets, S_1 and S_2, of four points in general position have a unique automorphism of $\mathrm{PGL}(2, \mathbf{F})$ transforming S_1 into S_2.

In the last two chapters we have tied these two approaches together, by showing that a Desarguesian projective plane is isomorphic to $\mathbf{P}^2(R)$ for a division ring

---

[1]cf. Section 5.2, Exercises 5.2, 5.7, 8.10.

we can construct, and conversely a projective plane $\mathbf{P}^2(R)$ satisfies Axiom P5. Additionally, we showed that Axioms P6 and P7 in our synthetic development are equivalent to algebraic statements about R in our analytic development on $\mathbf{P}^2(R)$. For example, every Pappian plane is of the form $\mathbf{P}^2(\mathbf{F})$, and is Fano iff char $\mathbf{F} \neq 2$.

In this chapter and the next, we continue to tie up our two approaches. Among the loose ends that are left are to give an analytic interpretation of the group of conjective projectivities $PJ(\ell)$, which we have so far studied only from the synthetic point of view. This is what we do in this chapter. In the next chapter, we give a synthetic interpretation of the subgroup $PGL(2, \mathbf{F})$ of automorphisms of a Pappian plane, which so far we have only studied from the analytic point of view. As a bonus of this interpretation, we prove Ceva's theorem, a basic theorem in advanced Euclidean geometry.

10.2   The Group of Möbius Transformations of the Extended Field

Let $\mathbf{F}$ be a field, and consider $\pi = \mathbf{P}^2(\mathbf{F})$, the projective plane over $\mathbf{F}$. Then π is a Pappian plane (which, you will recall, has Desargues' axiom).[2] Let ℓ be the line $x_3 = 0$: simplify the homogeneous coordinates for points on ℓ, $(x_1, x_2, 0)$, by writing just (x_1, x_2).

In our synthetic development in Chapters 5 and 6, we have studied the group $PJ(\ell)$ of conjective projectivities on ℓ. Now we will define another group $PGL(1, \mathbf{F})$ of transformations of ℓ into itself, and will prove it is equal to $PJ(\ell)$.

In the same spirit as Section 8.2, we define a transformation T_A of ℓ onto itself for each nonsingular matrix $A = \left(\begin{smallmatrix} a & b \\ c & d \end{smallmatrix}\right)$. Now it is a basic fact of linear algebra that nonsingularity of A is equivalent to the condition $\det A = ad - bc \neq 0$. Let $T_A(x_1, x_2) = (x_1', x_2')$, so that

$$\begin{aligned} x_1' &= ax_1 + bx_2 \\ x_2' &= cx_1 + dx_2, \end{aligned} \tag{10.1}$$

or in vector notation $T_A(x) = Ax = x'$.

Clearly, for each nonzero scalar λ in $\mathbf{F}$, $T_{\lambda A} = T_A$ since we are working in homogeneous coordinates: $T_{\lambda A}(x) = \lambda x' = x'$. Now if we wanted, we could work our way through a 2×2 variant of Section 8.2, proving that T_A is a one-to-one correspondence of ℓ with itself, whose inverse is $T_{A^{-1}}$ (Exercise 10.3). Moreover the set $\{T_X \colon \ell \to \ell \mid X \text{ is a } 2 \times 2 \text{ nonsingular matrix}\}$ forms a group (Exercise 10.4). Two matrices A and B define the same transformation, i.e. $T_A = T_B$, if and only if there is a nonzero $\lambda \in \mathbf{F}$ such that $B = \lambda A$ (Exercise 10.5). In addition, given three

---

[2]We stick to the commutative case for simplicity. Much of this section is valid over division rings — with a great deal of effort.

distinct points X_1, X_2, X_3 and another triple Y_1, Y_2, Y_3 we may find one and only one matrix system $\{\lambda A \mid \lambda \in \mathbf{F} \setminus \{0\}\}$ such that $T_A(X_i) = Y_i$ ($i = 1, 2, 3$): this you may do in Exercise 10.6. In analogy with Section 8.2, we have

10.1. Definition The group of transformations of ℓ into itself of the form T_A defined above, where A is a 2×2 nonsingular matrix over $\mathbf{F}$, is denoted by $\mathrm{PGL}(1, \mathbf{F})$.

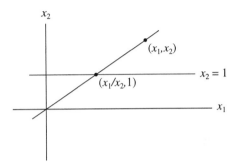

Figure 10.1. From homogeneous to inhomogeneous coordinates.

In carrying out our program of proving $\mathrm{PGL}(1, \mathbf{F}) = \mathrm{PJ}(\ell)$, we find it more convenient to introduce the *inhomogeneous coordinate* $x = x_1/x_2$ on ℓ. Note that x puts the points (x_1, x_2) of ℓ in one-to-one correspondence with the extended field $\mathbf{F} \cup \{\infty\}$, which we denote by $\mathbf{F}_\infty$: the notation ∞ corresponds to the single point $(x_1, 0)$, i.e. "$\infty = x_1/0$". (No special meaning should be given to ∞ here.)

By dividing the expressions for x_1' with that for x_2' in Equation 10.1 above it is apparent that the group $\mathrm{PGL}(1, \mathbf{F})$ is the set of transformations $T_A : x \mapsto x'$ where

$$x' = \frac{ax + b}{cx + d} \qquad (a, b, c, d \in \mathbf{F}, \ ad - bc \neq 0).$$

In addition $T_A(\infty) = \frac{a}{c}$ and $T_A(-\frac{d}{c}) = \infty$ if $c \neq 0$, and $T_A(\infty) = \infty$ if $c = 0$. Such transformations of the extended field are called *Möbius transformations* or *fractional linear transformations*, and recur throughout mathematics. You will recall seeing Möbius transformations of $\mathbf{C}_\infty$ in Section 6.2. The next proposition is proven just like Lemma 6.5.

10.2. Proposition *The set of Möbius transformations of* $\mathbf{F}_\infty$ *forms a group under composition: i.e.,* $\mathrm{PGL}(1, \mathbf{F})$ *is a group. Moreover, if* a, b, c *and* α, β, γ *are two triples of distinct elements of* $\mathbf{F}_\infty$, *there is a unique transformation* T_A *in* $\mathrm{PGL}(1, \mathbf{F})$ *which sends* a, b, c *into* α, β, γ, *respectively.*

Proof Exercise 10.8. □

10.3. Proposition *The group* $\mathrm{PGL}(1,\mathbf{F})$ *of Möbius transformations is generated by the sets* $I,\ \mathcal{J},\ \mathcal{K}$ *of Möbius transformations of three kinds:*

$$I = \left\{ T_A \,\middle|\, A = \begin{pmatrix} 1 & a \\ 0 & 1 \end{pmatrix},\ a \in \mathbf{F} \right\}$$

$$\mathcal{J} = \left\{ T_B \,\middle|\, B = \begin{pmatrix} a & 0 \\ 0 & 1 \end{pmatrix},\ a \in \mathbf{F} \smallsetminus \{0\} \right\}$$

$$\mathcal{K} = \left\{ T_C \,\middle|\, C = \begin{pmatrix} 0 & 1 \\ 1 & 0 \end{pmatrix} \right\}.$$

10.4. Remark In symbols, $\mathrm{PGL}(1,\mathbf{F}) = \langle I \cup \mathcal{J} \cup \mathcal{K} \rangle$. The transformations in I are translations with equation $x' = x + a$. The transformations in $\mathcal{J}$ are central dilatations. Sets I and $\mathcal{J}$ are in fact subgroups. The only transformation in $\mathcal{K}$ is called inversion, an order 2 element.

> **Proof** We start with a transformation $T(x) = \frac{ax+b}{cx+d}$ where $ad - bc \neq 0$. There are two cases to consider: $c = 0$ and $c \neq 0$.
> If $c = 0$, then $d \neq 0$. Let $T'_1(x) = \frac{a}{d}x$, so $T'_1 \in J$. Let $T'_2(x) = x + \frac{b}{d}$, so $T'_2 \in I$. Then $T'_2(T'_1(x)) = \frac{ax+b}{d} = T(x)$. We have disposed of this case.
> If $c \neq 0$, we can in fact work out that
>
> $$T(x) = (T_1 T_2 T_3 T_4)(x)$$
>
> where
>
> $$T_4(x) = x + \frac{d}{c}, \qquad T_3(x) = \frac{1}{x},$$
> $$T_2(x) = \frac{bc - ad}{c^2}x, \qquad T_1(x) = x + \frac{a}{c}.$$
>
> You will be asked to verify this in Exercise 10.4.
> Note that $T_1, T_4 \in I$, $T_2 \in J$ and $T_3 \in \mathcal{K}$, so I, $\mathcal{J}$, and $\mathcal{K}$ generate the group of Möbius transformations. □

10.3 $\mathrm{PJ}(\ell) \cong \mathrm{PGL}(1,\mathbf{F})$

10.5. Proposition *Each of the translations in I, central dilatations in $\mathcal{J}$ and the inversion in $\mathcal{K}$ in the Möbius transformations of $\mathbf{F}_\infty$ is a projectivity of ℓ ($x_3 = 0$) to itself. Hence,* $\mathrm{PGL}(1,\mathbf{F}) \subseteq \mathrm{PJ}(\ell)$.

> **Proof** We must exhibit each of the special Möbius transformations as a composition of perspectivities within $\mathbf{P}^2(\mathbf{F})$. Consider the line $x_2 = 0$ as the line

at ∞ of the affine plane with coordinates $x = x_1/x_2$ and $y = x_3/x_2$ (cf. Exercise 2.2 and Proposition 2.9). Then the line ℓ is $y = 0$, i.e. the x-axis.

Case *I*. We must show the transformation $T_I : x \mapsto x + a$ to be a projectivity. Consider

$$\ell \overset{(0,1)}{\underset{\wedge}{=}} \ell_\infty \overset{(a,1)}{\underset{\wedge}{=}} \ell.$$

This projectivity sends $(x,0)$ to the ideal point W lying on the line connecting $(0,1)$ and $(x,0)$, i.e. of slope $-\frac{1}{x}$. Now the line of slope $-\frac{1}{x}$ through $(a,1)$ intersects the x-axis, ℓ, at $(x+a,0)$. Hence x is sent to $x+a$ on ℓ, so T_I is equal to this projectivity.

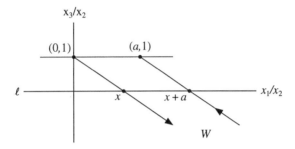

Figure 10.2. $x' = x + a$ as a projectivity.

Case *J*. Consider the transformation $T_J : x \mapsto ax$ $(a \neq 0)$. We claim T_J coincides on ℓ with the projectivity

$$\ell \overset{V}{\underset{\wedge}{=}} (x = y) \overset{W}{\underset{\wedge}{=}} \ell.$$

where V is ideal point on the vertical lines and W the ideal point on the line through $(1,1)$ and $(a,0)$. You will be asked to verify the details of T_J being this projectivity (Exercise 10.9a).

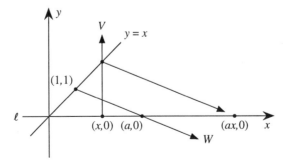

Figure 10.3. $x' = ax$ as a projectivity.

Case $\mathcal{K}$. The transformation $x' = \frac{1}{x}$ is a product of three perspectivities

$$\ell \overset{(1,1)}{\underset{\wedge}{=}} \ell_\infty \overset{(1,0)}{\underset{\wedge}{=}} (x=y) \overset{V}{\underset{\wedge}{=}} \ell$$

You will be asked to check that $(x,0)$ is sent to $(\frac{1}{x},0)$ in Exercise 10.9b.

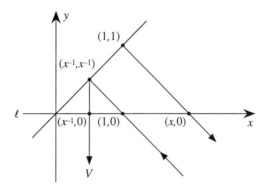

Figure 10.4. Inversion as a projectivity.

In conclusion, every Möbius transformation T is a projectivity of ℓ into itself, since T is a product of transformations in case I, $\mathcal{J}$, and $\mathcal{K}$ by Proposition 10.3.   □

10.6. Theorem *In a Pappian plane π with line ℓ, the group of projectivities of ℓ into itself is isomorphic to the group of Möbius transformations on the extended field $\mathbf{F}$ associated with π:*

$$PJ(\ell) \cong PGL(1,\mathbf{F}).$$

Proof We have seen in Proposition 10.5 that $PGL(1,\mathbf{F}) \subseteq PJ(\ell)$, where ℓ is the line $x_3 = 0$ in $\mathbf{P}^2(\mathbf{F})$. We establish the reverse inequality.

Let $\psi \in PJ(\ell)$, and suppose ψ takes the points 0, 1 and ∞ (or $(0,x_2)$, (x_1,x_1) and $(x_1,0)$ in homogeneous coordinates on ℓ) into X, Y, and Z, respectively (and projectively, of course). Now by Proposition 10.2, there exists a Möbius transformation T taking $0,1,\infty$ into X,Y,Z, respectively, and Proposition 10.5 establishes that T is a projectivity. By the fundamental theorem, $T = \psi$ since their values on three points agree. This completes our proof for the line ℓ.

Now given the lines ℓ and ℓ', in π it is a general fact that $PJ(\ell)$ and $PJ(\ell')$ are isomorphic as groups. For let ϕ be any perspectivity from ℓ onto ℓ'. Define a map $\psi\colon PJ(\ell) \to PJ(\ell')$ by

$$\alpha \mapsto \phi\alpha\phi^{-1}.$$

You will be asked to show in Exercise 10.6 that ψ is a group isomorphism. (Since ϕ is an arbitrary perspectivity, with no clear alternative, $PJ(\ell)$ and $PJ(\ell')$, though isomorphic, are not *canonically* so.) This shows the group of conjective projectivities to be an invariant of the projective plane, and removes the special nature of our computation with $x_3 = 0$. □

10.7. Remark Notice that our assumption of commutative scalars was put to full use in the form of the fundamental theorem. [Frankild-Kromann] investigates what is the case over a division ring R: Möbius transformations (suitably defined) still form a group, they are generated by translations, dilatations, and inversion, and we have $PGL(1,R) \cong PJ(\ell)$. However, the fundamental theorem in this case states that if ψ_1 and ψ_2 are two projectivities $ABC \overline{\wedge} A'B'C'$, then $\psi_1 \psi_2^{-1}$ is an inner automorphism of the division ring.

10.4 Cross Ratio: a Projective Invariant

We have seen in Exercise 5.7 that cross ratio is a projective invariant of $\mathbf{P}^2(\mathbf{R})$. This was done using special properties of the reals like the existence of trigonometric functions. Now cross ratio is clearly definable over any field $\mathbf{F}$ (cf. Exercise 8.10), since we need only subtract, divide and multiply in order to compute cross ratio. We will see in this section that cross ratio is also a projective invariant in $\mathbf{P}^2(\mathbf{F})$. Indeed the next theorem states that $PJ(\ell)$ is the group of permutations of points on ℓ that preserve cross ratio.

First, we give the definition of cross ratio in this more general setting. Let a, b, c, and d be four distinct points given in inhomogeneous coordinates for the line ℓ, $x_3 = 0$; i.e. $a,b,c,d \in \mathbf{F}_\infty$.

10.8. Definition The *cross ratio* is defined by

$$R_x(a,b;c,d) = \frac{a-c}{a-d} \cdot \frac{b-d}{b-c},$$

if none of $a,b,c,d = \infty$. In case one of a, b, c, or $d = \infty$ we set $R_x(a,b;c,d) = \frac{b-d}{b-c}$ if $a = \infty$, $\frac{a-c}{a-d}$ if $b = \infty$, $\frac{b-d}{a-d}$ if $c = \infty$, and $\frac{a-c}{b-c}$ if $d = \infty$.

10.9. Definition By a transformation of ℓ into itself that preserves cross ratio, we mean a one-to-one correspondence $\ell \to \ell$ sending each $X \in \ell$ into $X' \in \ell$ such that

$$R_x(A,B;C,D) = R_x(A',B';C',D')$$

for every quadruple of points A,B,C,D in ℓ. It is clear that the set of cross ratio preserving transformations of ℓ into itself is a group under composition of functions, which we denote by $\mathcal{R}(\ell)$.

10.10. Theorem *Let* **F** *be a field, and* ℓ *the line* $x_3 = 0$ *in* $\mathbf{P}^2(\mathbf{F})$. *Then the group of Möbius transformations (on the inhomogeneous coordinates* $\mathbf{F}_\infty$ *for* ℓ) *is equal to the group of permutations of* ℓ *that preserve cross ratio:*

$$\mathrm{PJ}(\ell) = \mathcal{R}(\ell).$$

Proof We first wish to prove that $\mathrm{PJ}(\ell) \subseteq \mathcal{R}(\ell)$. Now given a projectivity $T \colon \ell \barwedge \ell$ we have shown in Theorem 10.6 and its proof that T is a Möbius transformation of the inhomogeneous coordinates for ℓ. In Proposition 10.3 and its proof we saw how to factor T into a product of translations, central dilatations and inversions of $\mathbf{F}_\infty$.

It remains to show that $T \in \mathcal{R}(\ell)$ by showing that each of the three types of generating Möbius transformations preserve cross ratio

Case *I*. If $T(x) = x + \lambda$, translation by λ (and $T(\infty) = \infty$), we easily compute cross ratio of primed image points:

$$\mathrm{R}_x(a',b';c',d') = \frac{a+\lambda-(c+\lambda)}{a+\lambda-(d+\lambda)} \cdot \frac{b+\lambda-(d+\lambda)}{b+\lambda-(c+\lambda)},$$

which is clearly equal to $\mathrm{R}_x(a,b;c,d)$, also in the case that one point is ∞.

Case *J*. If $T(x) = \lambda x$ (and $T(\infty) = \infty$) where $\lambda \in \mathbf{F} \setminus \{0\}$, the transformed points satisfy

$$\mathrm{R}_x(a',b';c',d') = \frac{\lambda a - \lambda c}{\lambda a - \lambda d} \cdot \frac{\lambda b - \lambda d}{\lambda b - \lambda c},$$

which again is clearly equal to $\mathrm{R}_x(a,b;c,d)$, also in case one point is ∞.

Case *K*. If $T(x) = \frac{1}{x}$ ($T(\infty) = 0$, $T(0) = \infty$) we compute cross ratio of $T(a) = a'$, etc., to be

$$\mathrm{R}_x(a',b';c',d') = \frac{\frac{1}{a} - \frac{1}{c}}{\frac{1}{a} - \frac{1}{d}} \cdot \frac{\frac{1}{b} - \frac{1}{d}}{\frac{1}{b} - \frac{1}{c}} \cdot \frac{abcd}{abcd}$$

$$= \frac{c-a}{d-a} \cdot \frac{d-b}{c-b} = \mathrm{R}_x(a,b;c,d)$$

if none of a, b, c, or $d = \infty$ or 0. If $a = \infty$,

$$\mathrm{R}_x(a',b';c',d') = \frac{b}{b} \cdot \frac{d}{c} \cdot \frac{\frac{1}{b} - \frac{1}{d}}{\frac{1}{b} - \frac{1}{c}} = \frac{d-b}{c-b} = \mathrm{R}_x(\infty,b;c,d).$$

If $a = 0$,

$$\mathrm{R}_x(a',b';c',d') = \frac{bcd}{bcd} \cdot \frac{\frac{1}{b} - \frac{1}{d}}{\frac{1}{b} - \frac{1}{c}} = \frac{c}{d} \cdot \frac{d-b}{c-b} = \mathrm{R}_x(0,b;c,d).$$

The other cases proceed similarly (Exercise 10.10).

This establishes that $\mathrm{PJ}(\ell) \subseteq \mathcal{R}(\ell)$.

Conversely, suppose $\phi \in \mathcal{R}(\ell)$. Let $a = \phi(0)$, $b = \phi(1)$, $c = \phi(\infty)$ and generally $x' = \phi(x)$. By hypothesis $R_x(a,b;c,x') = R_x(0,1;\infty,x)$, i.e.

$$\frac{a-c}{a-x'} \cdot \frac{b-x'}{b-c} = \frac{1-x}{-x}.$$

Suppose for the moment that $a,b,c \neq \infty$. Upon solving for x' in three steps we have

$$x(a-c)(b-x') = (x-1)(a-x')(b-c)$$
$$x(a-c)b + x(c-b)a + a(b-c) = axx' - bxx' + (b-c)x'$$
$$x' = \frac{\frac{a-b}{b-c}cx + a}{\frac{a-b}{b-c}x + 1}$$

Since

$$\frac{a-b}{b-c}c - a\frac{a-b}{b-c} = \frac{(c-a)(a-b)}{b-c} \neq 0$$

as **F**-elements, we conclude that ϕ is a Möbius transformation. If $a = \infty = \phi(0)$, we get $\frac{b-x'}{b-c} = \frac{x-1}{x}$, so

$$x' = -(b-c)\frac{x-1}{x} + b = \frac{cx + c - b}{x},$$

which is also a Möbius transformation. You will be asked to check the other cases in Exercise 10.10.

Hence, $\phi \in \mathrm{PJ}(\ell)$ and so $\mathrm{PJ}(\ell) = \mathcal{R}(\ell)$. $\square$

10.11. Remark In the spirit of calling $\mathbf{F}^{n+1}/\sim$ projective n-space over **F**, where $(x_1,\ldots,x_{n+1}) \sim (x_1\lambda,\ldots,x_{n+1}\lambda)$, we call $\mathbf{F}^2/\sim$ the *projective line over* **F**, denoted by $\mathbf{P}^1(\mathbf{F})$. We saw that the line ℓ with inhomogeneous coordinates could be viewed as the extended "number line" $\mathbf{F}_\infty$. Now there is nothing special about $x_3 = 0$: in the first instance, we see that $x_2 = 0$ or $x_1 = 0$ may replace ℓ and still be put in a canonical one-to-one correspondence with $\mathbf{F}_\infty$. In fact, any line m may be transformed to ℓ by a projective collineation, defined in the next chapter, which puts m in one-to-one correspondence with $\mathbf{F}_\infty$ (only up to a Möbius transformation, since there will be more than one projective collineation transforming m into ℓ). This has two consequences. One is that any line in $\mathbf{P}^2(\mathbf{F})$ is projectively equivalent to the projective line over **F**. The other is that cross ratio is definable for every line and the last theorem, Theorem 10.10, is valid for any line in $\mathbf{P}^2(\mathbf{F})$.

In case $\mathbf{F} = \mathbf{R}$ or $\mathbf{C}$, the projective line over $\mathbf{F}$ has certain well-known topological models. The projective line over the reals is a circle and the projective line over the complex numbers is a sphere (the *Riemann sphere*), both of which can be seen by stereographic projection in dimension 2 and 3, respectively, from the "north pole", which corresponds to ∞ under this mapping.

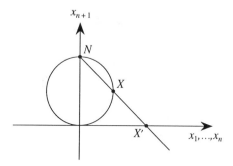

Figure 10.5. Stereographic projection of the n-sphere S^n into $\mathbf{R}^n \cup \{\infty\}$.

These identifications are useful for putting functions on the circle in one-to-one correspondence with functions of two homogeneous variables.

It is shown in a course in complex analysis that the group of Möbius transformations of the Riemann sphere is equal to the group of orientation-preserving, conformal (i.e. angle-preserving) one-to-one transformations of the Riemann sphere onto itself. These are automatically infinitely differentiable at all points except one, where a "simple pole" takes place. This provides a fourth interpretation of the same group $\mathrm{PJ}(\ell)$ in the case $\mathbf{F} = \mathbf{C}$.

Exercises

10.1   Consider the line $\ell \colon x_3 = 0$ in the seven point plane $\mathbf{P}^2(\mathbf{Z}_2)$. List the Möbius transformations in $\mathrm{PGL}(1, \mathbf{Z}_2)$, and compare with your computation of $\mathrm{PJ}(\ell)$ in Exercise 5.6.

10.2   Assume $\mathbf{F}$ is a field of characteristic different from 2 or 3. Determine the Möbius transformation of $\mathbf{F}_\infty$ which sends $0 \mapsto -1$, $2 \mapsto 3$, and $3 \mapsto 2$. Write the transformation you find as a composition of translations, stretchings and inversions.

Exercises 10.3–10.6 provide an alternative demonstration of Proposition 10.2, and may be done by adapting the proofs of Propositions 8.8, 8.12 and Theorem 8.13 to 2×2 nonsingular matrices over a field $\mathbf{F}$.

10.3   Let ℓ be the line $x_3 = 0$, A a nonsingular 2×2 matrix, and T_A the transformation $x' = Ax$. Show that T_A is a one-to-one correspondence of ℓ into itself, with inverse given by $T_{A^{-1}}$.

10.4   Show that the set $\{T_A : \ell \to \ell \mid \det A \neq 0\}$ is a group under composition.

10.5   Show that $T_A = T_B : \ell \to \ell$ if and only if there exists $\lambda \in \mathbf{F} \setminus \{0\}$ such that $A = \lambda B$.

10.6   Let X_1, X_2, X_3 and Y_1, Y_2, Y_3 be two triples of points on ℓ, no two of which are equal. Show that there is a matrix A, unique up to a scalar, such that $T_A(X_i) = Y_i$ for $i = 1$, 2, and 3.

10.7   In Section 8.2 we chose to work with $\mathrm{PGL}(2, \mathbf{R})$ using homogeneous coordinates and 3×3 nonsingular matrices. This is in direct contrast to our use of inhomogeneous coordinates $\mathbf{F}_\infty$ and Möbius transformations in Section 10.1. In this exercise you will see what you obtain if you pass to inhomogeneous coordinates in a study of $\mathrm{PGL}(2, \mathbf{F})$ for some field $\mathbf{F}$.

     (a) Let $T_A \in \mathrm{PGL}(2, \mathbf{F})$ where $A = (a_{ij})$ and $\det A \neq 0$. Show that T_A is given by

$$x' = \frac{a_{11}x + a_{12}y + a_{13}}{a_{31}x + a_{32}y + a_{33}}, \qquad y' = \frac{a_{21}x + a_{22}y + a_{23}}{a_{31}x + a_{32}y + a_{33}}$$

in the inhomogeneous coordinates $x = x_1/x_3$, $y = x_2/x_3$.

     (b) Show that the ideal point on the affine plane $x_3 = 1$ corresponding to slope m is sent to

$$x' = \frac{a_{11} + ma_{12}}{a_{31} + ma_{32}}, \qquad y' = \frac{a_{21} + ma_{22}}{a_{31} + ma_{32}}.$$

How is this to be interpreted if $m = \infty$?

10.8   Prove that $\mathrm{PGL}(1, \mathbf{F})$ is a 3-transitive group: refer to the statement of Proposition 10.2 and the proof of Lemma 6.5.

10.9   Let ℓ be the line $x_3 = 0$ in $\mathbf{P}^2(\mathbf{F})$, which is made to correspond to the x-axis in the affine coordinates $y = x_3/x_2$ and $x = x_1/x_2$.

     (a) Consider the projectivity

$$\ell \; \overset{V}{\underset{\wedge}{=}} \; (x = y) \; \overset{W}{\underset{\wedge}{=}} \; \ell$$

as in the proof of Proposition 10.5, where V is the ideal point on the y-axis and W is the ideal point on the line through $(1, 1)$ and $(a, 0)$ where $a \neq 0$. Show that $(x, 0)$ is sent into $(ax, 0)$.

     (b) Consider the projectivity

$$\ell \; \overset{(1,1)}{\underset{\wedge}{=}} \; \ell_\infty \; \overset{(1,0)}{\underset{\wedge}{=}} \; (x = y) \; \overset{V}{\underset{\wedge}{=}} \; \ell,$$

where V is the same as in (a). Show that the projectivity sends $(x, 0)$ into $(\frac{1}{x}, 0)$.

10.10   Let ϕ be a cross ratio preserving transformation of $\mathbf{F}_\infty$. Let $a = \phi(0)$, $b = \phi(1)$, and $c = \phi(\infty)$. Suppose that one of b or c is ∞. Show that $\phi(x) = x'$ is a Möbius transformation.

10.11   A linear transformation τ of a vector space V is called an *involution* if $\tau^2 = \mathrm{id}_V$. Suppose V is a 2-dimensional vector space over a field $\mathbf{F}$. Show that every invertible linear transformation $T : V \to V$ is, up to a scalar multiple, either an involution or is the product of two involutions.

     Hint: What does Exercise 6.7 say for linear transformations?

10.12   Show that the set of Möbius transformations of $\mathbf{F}_\infty$ can be mapped injectively to a subset of $\mathbf{P}^3(\mathbf{F})$.

10.13   Suppose $f\colon \mathbf{F}_\infty \to \mathbf{F}_\infty$ sends harmonic quadruples into harmonic quadruples. Moreover, suppose $f(0) = 0$, $f(1) = 1$ and $f(\infty) = \infty$. Prove that f restricted to $\mathbf{F}$ is a field automorphism.

10.14   Suppose $\mathbf{F}$ and $\mathbf{K}$ are fields, and $\sigma\colon \mathbf{P}^1(\mathbf{F}) \to \mathbf{P}^1(\mathbf{K})$ is a bijection between projective lines which sends harmonic points into harmonic points.

   (*a*) Choose a Möbius transformation τ of $\mathbf{P}^1(\mathbf{K})$ such that $\eta = \tau \circ \sigma$ satisfies $\eta(0) = 0$, $\eta(1) = 1$, and $\eta(\infty) = \infty$.

   (*b*) von Staudt: By using certain well-chosen harmonic quadruples and Exercise 7.22 show that $\eta|_{\mathbf{F}}$ is an isomorphism of the field $\mathbf{F}$ onto the field $\mathbf{K}$: i.e., η is bijective, $\eta(x+y) = \eta(x) + \eta(y)$, and $\eta(xy) = \eta(x)\eta(y)$ for all $x, y \in \mathbf{F}$.

   (*c*) G. Ancochea, Hua: Now suppose that $\mathbf{F}$ and $\mathbf{K}$ are division rings and that $\sigma\colon \mathbf{P}^1(\mathbf{F}) \to \mathbf{P}^1(\mathbf{K})$ is a harmonic point preserving bijection. Show that there exists a projectivity $\tau\colon \mathbf{P}^1(\mathbf{K}) \to \mathbf{P}^1(\mathbf{F})$ and either an isomorphism or anti-isomorphism $\eta\colon \mathbf{F} \to \mathbf{K}$ such that $\sigma = \tau \circ \eta$. (An *anti-isomorphism* $\eta\colon \mathbf{F} \to \mathbf{K}$ of rings is an additive bijective map that satisfies $\eta(xy) = \eta(y)\eta(x)$ for all $x, y \in \mathbf{F}$.)

10.15   Let π be a Pappian plane, A and B two points in π, and ℓ the line AB. Show that the mapping f_{AB} defined by

$$f_{AB}(C) = \text{harmonic conjugate of } C \text{ w.r.t. } A, B.$$

is a projectivity $\ell \barwedge \ell$. How should f_{AB} be defined on A and B?

Chapter 11

Projective Collineations

In this chapter, we develop the synthetic theory of projective collineation. In general, *collineation* is a synonym for automorphism of a projective plane, because lines are sent into lines.

11.1 Projective Collineations

11.1. Definition Let ℓ be a line of the projective plane π. A *projective collineation* is an automorphism $\phi \colon \pi \to \pi$, $\ell \mapsto \ell'$ such that ϕ restricted to ℓ is a projectivity

$$\phi|_\ell \colon \ell \overline{\wedge} \ell'.$$

Now this definition will probably seem vague on a first reading, for could not ϕ be a projectivity on ℓ but fail to be on some other line? We will see in the next proposition that this is not the case.

11.2. Proposition *Let ϕ be a projective collineation of π. Then, for any line m, $\phi|_m$ is a projectivity.*

> **Proof** Assuming $\phi|_\ell$ is a projectivity, $\ell \overline{\wedge} \ell'$, we wish to show that $\phi|_m \colon m \to m'$ is a projectivity as well. Let P be a point not on ℓ or m.
>
> Let $\rho \colon m \to \ell$ be the perspectivity $m \overset{P}{\overline{\wedge}} \ell$. Let $B \in m$ and $\rho(B) = A$. Then A, B and P are collinear, and so are their image points, A', B' and P' under our automorphism ϕ. Clearly, $A' \in \ell'$ and $B' \in m'$.

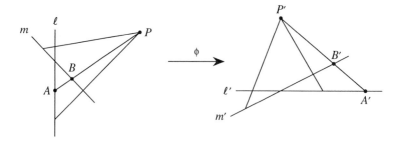

Figure 11.1. $\phi|_m$ is a projectivity.

Let $\tau\colon \ell' \mapsto m'$ be the perspectivity with center P'. We compare the mappings $\phi|_m$ and $\tau \circ \phi|_\ell \circ \rho$, both of which send m into m'. On the one hand, $\phi|_m(B) = B'$; on the other hand, $\tau \circ \phi|_\ell \circ \rho(B) = \tau \circ \phi(A) = \tau(A') = B'$. Since B is arbitrary, we have

$$\phi|_m = \tau \circ \phi|_\ell \circ \rho.$$

But the right-hand mapping is a projectivity. $\qquad\qquad\qquad\qquad\square$

The simplest example of a projective collineation is the identity. We will next study two key examples of projective collineations, called elations and homologies. We will prove that if π is a Pappian plane, then any projective collineation is a composition of at most three elations and two homologies. Finally, we will be able to show that if $\pi \cong \mathbf{P}^2(\mathbf{F})$ where $\mathbf{F}$ is a field, then the group of projective collineations is precisely $\mathrm{PGL}(2,\mathbf{F})$.

For now, let us apply our knowledge gained from Chapter 10 in order to note an example of an automorphism that is not a projective collineation. For this it suffices to find an automorphism that fails to preserve cross ratio, by Theorem 10.10. From Chapter 8 we know that an automorphism σ of $\mathbf{F}$ gives an automorphism S_σ of $\mathbf{P}^2(\mathbf{F})$. Since complex conjugation, $z \mapsto \bar{z}$, does not preserve cross ratio, complex conjugation will induce an automorphism of $\mathbf{P}^2(\mathbf{C})$ that is *not* a projective collineation (Exercise 11.1).

11.2 Elations and Homologies

It follows from the fact that projective collineations transform every line projectively that the set of projective collineations form a group, denoted by $\mathrm{PC}(\pi)$, within the group of automorphisms of a projective plane π. As promised, we turn to the study of elations and homologies.[1]

---

[1] This terminology is due to Sophus Lie (1842–1899).

11.3. Definition An *elation* is an automorphism α of π that leaves fixed each point P of some line ℓ, called the *axis*, but fixes no other points of π:

$$\alpha(P) = P \qquad \Longleftrightarrow \qquad P \in \ell.$$

We have seen something like an elation before. If we remove from π the line ℓ, we are left with an affine plane **A** as in Exercise 2.2. We will now show that $\alpha|_\mathbf{A}$ is a translation different from the identity. Indeed we need only show $\alpha|_\mathbf{A}$ is a dilatation since α has no fixed points outside of the removed line ℓ. Then given $P, Q \in \mathbf{A}$ we must show that the transformed points P', Q' lie on a line parallel to PQ. Suppose PQ intersects ℓ in the point W in π. Then P, Q and W are collinear. Since $\alpha(W) = W$, it follows that P', Q' and W are collinear. Since pencils of parallels in **A** are pencils of lines on points of ℓ, we arrive at $PQ \parallel P'Q'$. Thus an elation restricts to a dilatation of **A**: indeed, $\alpha|_\mathbf{A}$ is a translation since it has no fixed points.

Conversely, it is easy to see that a translation gives an elation α of the completed plane by defining $\alpha(X) = X$ on each ideal point X (Exercise 11.2).

11.4. Proposition *Let π be any projective plane. The elations with axis ℓ, together with the identity transformation, form a group $\mathcal{E}(\ell)$ under composition, itself a subgroup of* $\mathrm{PC}(\pi)$. *Also,* $\mathcal{E}(\ell) \cong \mathrm{Tran}\,\mathbf{A}$.

> **Proof** Since an elation is the identity projectivity on its axis, it is a projective collineation. It is easy to see that if $\alpha, \beta \in \mathcal{E}(\ell)$, then $\alpha \circ \beta^{-1} \in \mathcal{E}(\ell)$, so $\mathcal{E}(\ell)$ is a group (Exercise 11.3). Now the foregoing discussion has shown how to associate a translation to each elation, and vice versa. This bijection is in fact a group isomorphism (Exercise 11.4). $\qquad \square$

If α is an elation with axis ℓ, then we have noted that $\alpha|_\mathbf{A}$ is a translation. Recall that for any $P, Q \in \mathbf{A}$, $PP' \parallel QQ'$. Let $PP'.\ell = X$. We call X the *center* of the elation α. In the real affine plane, X would be the direction of the translation $\alpha|_\mathbf{A}$.

Although $\mathcal{E}(\ell)$ is a group, one should not suppose that all elations taken together form a group. For if α and β are elations with different axes, ℓ and m, there is no reason to suppose that $\alpha\beta$ is an elation. Exercise 11.5 asks you to find a "counterexample" in the 7 point plane.

However, there is something we can say about all elations. It will turn out that, with the addition of Axiom P5, the two subgroups $\mathcal{E}(\ell)$ and $\mathcal{E}(m)$ of $\mathrm{Aut}(\pi)$ are isomorphic in a very special way: they are conjugate subgroups, a group-theoretic concept we now define.

11.5. Definition Let G be a group. Let H and K be subgroups of G. H and K are called *conjugate* subgroups, if there is an element $g \in G$ such that the map

$$h \mapsto ghg^{-1}$$

is an isomorphism of H onto K. That the mapping is a group isomorphism is a routine exercise (Exercises 11.6). This is sometimes denoted by $K = gHg^{-1}$. Note that symmetrically one has $H = g^{-1}Kg$.

11.6. Proposition *Let π be a Desarguesian plane. Then the groups of elations $\mathcal{E}(\ell)$ and $\mathcal{E}(m)$ are conjugate subgroups in $\mathrm{Aut}(\pi)$.*

> **Proof** We pick an automorphism ϕ that sends ℓ into m (cf. Theorem 8.13). Then the mapping
>
> $$\alpha \mapsto \phi \circ \alpha \circ \phi^{-1} \qquad (\alpha \in \mathcal{E}(\ell))$$
>
> is an isomorphism of $\mathcal{E}(\ell)$ onto $\mathcal{E}(m)$. Indeed, $\phi\alpha\phi^{-1}$ fixes P if and only if $P \in m$, so $\phi\alpha\phi^{-1} \in \mathcal{E}(m)$ (Exercise 11.7). □

We now turn to the other type of projective collineation — homology — which turns out to be closely related to central dilatation of the affine plane.

11.7. Definition A *homology* of the projective plane π is a projective collineation α of π leaving a line ℓ pointwise fixed and fixing precisely one other point O in $\pi - \ell$. α is said to be the homology with *axis* ℓ and *center* O. The set of homologies with axis ℓ and center O is a group denoted by $\mathcal{H}(\ell, O)$. The set of homologies together with elations, all with axis ℓ but of arbitrary center on or off ℓ, is a group denoted by $\mathcal{H}(\ell)$.

11.8. Proposition *Let π be a Desarguesian plane. Then $\mathcal{H}(\ell)$ is a semi-direct product of its subgroups $\mathcal{E}(\ell)$ and $\mathcal{H}(\ell, O)$.*

> **Proof** Let $\pi \smallsetminus \ell$ be the affine plane **A**. **A** satisfies the major and minor Desargues' axioms as noted in Chapter 9. The mapping
>
> $$\alpha \mapsto \alpha|_{\mathbf{A}}$$
>
> is an isomorphism of $\mathcal{H}(\ell)$ onto $\mathrm{Dil}\,\mathbf{A}$ sending $\mathcal{H}(\ell, O)$ onto the group of central dilatations $\mathrm{Dil}_O(\mathbf{A})$ and sending $\mathcal{E}(\ell)$ onto $\mathrm{Tran}\,\mathbf{A}$: this claim follows from Proposition 11.4 and the preceding discussion. In Proposition 9.11 it was shown that $\mathrm{Dil}\,\mathbf{A}$ is the semi-direct product of the normal, abelian subgroup $\mathrm{Tran}\,\mathbf{A}$ and any one of its subgroups $\mathrm{Dil}_O(\mathbf{A})$ for $O \in \mathbf{A}$. But group isomorphism preserves a semi-direct product structure. You will be asked to check the details in Exercise 11.8. □

11.3   The Fundamental Theorem of Projective Collineation

The next proposition, aside from its intrinsic interest, is a key step in the proof of Theorem 11.11.

11.9. Proposition *Let π be a Desarguesian plane. Let A,B,C,D and A',B',C',D' be two quadruples of points, no three of which are collinear. Then one can find a product ϕ of elations and homologies such that $\phi(A) = A'$, $\phi(B) = B'$, $\phi(C) = C'$ and $\phi(D) = D'$.*

> **Proof** The proof proceeds in five steps as we show that, in general, ϕ is a product of three elations and two homologies.
>
>   *(1)* Choose a line ℓ not incident with either A or A'. Since π is Desarguesian, it follows from Theorem 9.6 that there is a translation of the affine plane $\pi \smallsetminus \ell$ which sends A into A', and as we have seen in the last section this gives an elation $\alpha_1 : \pi \to \pi$ with axis ℓ such that $\alpha_1(A) = A'$. Denote $\alpha_1(B) = B_1$, $\alpha_1(C) = C_1$ and $\alpha_1(D) = D_1$.
>
>   *(2)* Since α_1 will be the first in a product of elations and homologies, we now wish to fix A' and send B_1 into B' as a second step. We can do this by choosing a line m incident with A' but not incident with either B_1 or B'.
>
>   As before there exists a unique elation α_2 with axis m such that $\alpha_2(B_1) = B'$. Note that $\alpha_2(A') = A'$ since A' is on the axis. Denote $\alpha_2(C_1) = C_2$ and $\alpha_2(D_1) = D_2$.
>
>   *(3)* This time apply Theorem 9.6 to transform C_2 into C' by an elation α_3 with axis $A'B'$. Note that there is no problem with this, since $C' \notin A'B'$ and $C \notin AB$ by hypothesis, so
>
> $$\alpha_2\alpha_1(C) = C_2 \notin \alpha_2\alpha_1(A) \cup \alpha_2\alpha_1(B) = A'B',$$
>
> since α_1, α_2 are collineations. So far $\alpha_3 \circ \alpha_2 \circ \alpha_1$ transforms A, B, C into A', B', C', respectively and, denoting $\alpha_3(D_2) = D_3$, D transforms into D_3.
>
>   *(4)* In the last two steps we will fix A', B', C' and transform D_3 by two homologies into D'.
>
>   Let $D_4 = A'D_3.B'D'$. Now in the affine plane $\pi \smallsetminus B'C'$, in which the major Desargues' axiom holds, we find by Theorem 9.12 a (unique) central dilatation $\hat{\beta}_4$ holding A' fixed and sending D_3 into D_4, for A', D_3 and D_4 are collinear points, none of which lie on $B'C'$ (why?). So if β_4 denotes the homology with axis $B'C'$ and center A' corresponding to $\hat{\beta}_4$, then $\beta_4(A') = A'$, $\beta_4(B') = B'$, $\beta_4(C') = C'$ and $\beta_4(D_3) = D_4$.

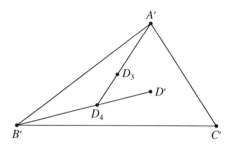

Figure 11.2. Transforming D_3 into D' in two steps.

(5) This time apply Theorem 9.12 to get an homology β_5 with axis $A'C'$, center B', such that $\beta_5(D_4) = D'$. There is no problem doing this, because B', D_4 and D' are collinear points, none of which lie on $A'C'$ (why?).

In conclusion, $\phi = \beta_5\beta_4\alpha_3\alpha_2\alpha_1$ transforms A, B, C, D into A', B', C', D', respectively.   □

11.10. Lemma *Let π be a Pappian plane. Let ϕ be a projective collineation of π, which leaves fixed four points A, B, C and D, no three of which are collinear. Then ϕ is the identity transformation.*

Proof Let ℓ denote the line BC. Since $\phi(B) = B$ and $\phi(C) = C$, ϕ sends ℓ into itself. Indeed, $\phi|_\ell$ is a projectivity of ℓ into itself, since ϕ is a projective collineation.

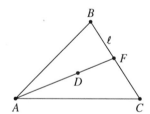

Figure 11.3. Three fixed points on a line.

Since ϕ also fixes A and D, it follows that $F = AD.\ell$ is fixed by ϕ. Whence, $\phi|_\ell : BFC \overline{\wedge} BFC$, so $\phi|_\ell = \text{id}_\ell$ by the fundamental theorem (of Chapter 6).

Let **A** denote the affine plane $\pi \smallsetminus \ell$. Since ϕ fixes each point on ℓ, we have seen in the discussion preceding Proposition 11.4 that $\phi|_{\mathbf{A}}$ is a dilatation. But $\phi|_{\mathbf{A}}$ has the two fixed points A and D. Hence, $\phi|_{\mathbf{A}} = \text{id}_{\mathbf{A}}$ by Proposition 1.19. Putting $\phi|_{\mathbf{A}} = \text{id}_{\mathbf{A}}$ and $\phi|_\ell = \text{id}_\ell$ together, we conclude that $\phi = \text{id}_\pi$.   □

We now come to the fundamental theorem of projective collineation.

11.11. Fundamental theorem of projective collineation *Let π be a Pappian plane. If A,B,C,D and A',B',C',D' are two quadruples of points, no three of which are collinear, then there is a unique projective collineation ϕ such that*

$$\phi(A) = A', \quad \phi(B) = B', \quad \phi(C) = C', \quad \phi(D) = D'.$$

Moreover, the group PC(π) is generated by elations and homologies.

Proof In Proposition 11.9 we proved that ϕ could be chosen to be a product of elations and homologies. Since each elation and homology is a projective collineation, it follows that $\phi \in$ PC(π).

We are left with proving uniqueness. Suppose ϕ, ψ are two projective collineations transforming A,B,C,D into A',B',C',D', respectively. Then $\phi \circ \psi^{-1} \in$ PC(π), and $\phi \circ \psi^{-1}$ fixes A', B', C' and D'. We may apply Lemma 11.10 to get $\phi \circ \psi^{-1} = \text{id}_{\pi}$, so $\phi = \psi$, which gives uniqueness.

Finally, it is clear that given $\eta \in$ PC(π), η will coincide with a product of elations and homologies on four points in general position. By an application of the previous lemma, η is then the product of elations and homologies. Hence, the union of the subset of elations with the subset of homologies generates PC(π). $\square$

11.4   PC$(\pi) \cong$ PGL$(2,\mathbf{F})$

At this point, we can realize our program of giving a synthetic interpretation to PGL$(2,\mathbf{F})$.

11.12. Theorem *Let $\mathbf{F}$ be a field, and let $\pi = \mathbf{P}^2(\mathbf{F})$ be the projective plane over $\mathbf{F}$. Then*

$$\text{PC}(\pi) = \text{PGL}(2,\mathbf{F}).$$

Proof We first show that elations and homologies can be given by matrices.

Let ℓ be the line $x_3 = 0$, and P the point $(1,0,0)$ on ℓ. Then $\pi \smallsetminus \ell$ is the affine plane $\mathbf{A}$ consisting of points with coordinates $x_3 \neq 0$. The affine coordinates of $\mathbf{A}$ are as usual

$$x = x_1/x_3$$
$$y = x_2/x_3.$$

Now an elation of π with axis ℓ and center P is a translation of $\mathbf{A}$, and conversely. Perhaps the simplest example of a translation $\alpha|_{\mathbf{A}}$ is ($a \in \mathbf{F}$)

$$x' = x + a$$
$$y' = y,$$

with homogeneous coordinates

$$\begin{aligned} x_1' &= x_1 + ax_3 \\ x_2' &= x_2 \\ x_3' &= x_3. \end{aligned}$$

Hence the elation $\alpha \colon \pi \to \pi$ is represented by the matrix

$$A = \begin{pmatrix} 1 & 0 & a \\ 0 & 1 & 0 \\ 0 & 0 & 1 \end{pmatrix}$$

with $a \in \mathbf{F}$; i.e. $x' = \alpha(x) = Ax$ is the equation defining α. Thus $\alpha \in \mathrm{PGL}(2, \mathbf{F})$.

If β is any other elation, with axis m and center Q, we can always find a nonsingular 3×3 matrix X such that the induced automorphism T_X of π sends ℓ into m and P into Q (cf. Theorem 8.13). Then β and $T_X \circ \alpha \circ T_X^{-1}$ are two automorphisms both fixing each point on line m. Now the center of $T_X \circ \alpha \circ T_X^{-1}$ may be computed as follows: since

$$T_X^{-1}(R) \cup \alpha(T_X^{-1}(R)).\ell = P,$$

it follows by applying T_X to both sides of the incidence equation that

$$R \cup T_X \alpha T_X^{-1}(R).m = Q.$$

Hence, $T_X \circ \alpha \circ T_X^{-1}$ has center Q, so $T_X \circ \alpha \circ T_X^{-1} = \beta$. But $\alpha = T_A$, so $\beta = T_{XAX^{-1}}$ (i.e., β is represented by XAX^{-1} with X a transition matrix). Hence, each elation of π is representable by a matrix that may be obtained from A by a change of basis.

Thus every elation is in $\mathrm{PGL}(2, \mathbf{F})$. Now consider a homology γ with axis $x_1 = 0$ and center $(1, 0, 0)$ off the axis. In the affine plane $x_1 \neq 0$ with affine coordinates $x = x_2/x_1$, $y = x_3/x_1$, $\gamma|_{\mathbf{A}}$ is a central dilatation with center $(0, 0)$, hence a stretching in some ratio $k \neq 0$. It will have equation in homogeneous coordinates

$$\begin{aligned} x_1' &= x_1 \\ x_2' &= kx_2 \\ x_3' &= kx_3, \end{aligned}$$

i.e. $\gamma = T_C$ where

$$C = \begin{pmatrix} 1 & 0 & 0 \\ 0 & k & 0 \\ 0 & 0 & k \end{pmatrix}.$$

As before, any other homology δ has a matrix similar to C, i.e. there exists $Y \in \mathrm{GL}_3(\mathbf{F})$ such that $\delta = T_{YCY^{-1}}$ (Exercise 11.9).

Hence, every elation and homology, and indeed every product of such, are of the form T_A for some 3×3 nonsingular matrix A over $\mathbf{F}$. So by Theorem 11.11,

$$\mathrm{PC}(\pi) \subseteq \mathrm{PGL}(2,\mathbf{F}).$$

We finish in what is by now a standard way.[2] Let $T \in \mathrm{PGL}(2,\mathbf{F})$. Let A,B,C,D be a quadruple of points in general position, and suppose $T(A) = A'$, $T(B) = B'$, $T(C) = C'$ and $T(D) = D'$. Now A',B',C',D' is also a quadruple in general position since T is a collineation. By the fundamental theorem of projective collineation, there exists a unique $\beta \in \mathrm{PC}(\pi)$ such that $\beta(A) = A'$, $\beta(B) = B'$, $\beta(C) = C'$ and $\beta(D) = D'$. But we have shown in the paragraphs above that $\beta \in \mathrm{PGL}(2,\mathbf{F})$. So we have β, T in $\mathrm{PGL}(2,\mathbf{F})$, having the same values on the quadruple A,B,C,D in general position, so by Theorem 8.13 we have $\beta = T$ since $\mathbf{F}$ is a field. So we have the reverse inclusion

$$\mathrm{PGL}(2,\mathbf{F}) \subseteq \mathrm{PC}(\pi).$$

$\square$

If we started with an isomorphism of projective planes $\pi \cong \mathbf{P}^2(\mathbf{F})$, we would only arrive at an isomorphism of groups $\mathrm{PC}(\pi) \cong \mathrm{PGL}(2,\mathbf{F})$. Why?

11.5   Ceva's Theorem

In this section we prove an important theorem in advanced Euclidean geometry known as Ceva's theorem.[3] We will make full use of our projective theory of cross ratio and projective collineation in order to prove a generalization of Ceva's theorem to Pappian planes. In another direction, the interested reader can see an application of our theory of projective collineation to matrices and determinants in Exercise 11.10.

11.13. Theorem *Let π be the Pappian, Fano plane $\pi = \mathbf{P}^2(\mathbf{F})$, $\mathbf{F}$ a field of characteristic $\neq 2$. Let ABC be a triangle in π, and ℓ a line different from AB, BC, or AC. Denote the points of intersection by*

$$U = \ell.AB, \quad V = \ell.BC, \quad W = \ell.CA.$$

Let E, F and G be arbitrary points on BC, AB, and AC, respectively, each different from A,B,C. Then the lines BG, AE and CF are concurrent if and only if the cross ratios satisfy

$$\mathrm{R}_x(A,B;F,U)\mathrm{R}_x(B,C;E,V)\mathrm{R}_x(C,A;G,W) = -1. \tag{11.1}$$

---

[2] You might cover the rest of the paragraph and try it as an exercise.

[3] After Giovanni Ceva (c. 1647–1736). If we are given a triangle *ABC* with a point *D* in general position, the lines *AD*, *BD*, and *CD* are named *cevians* in his honor.

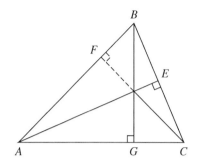

Figure 11.5. The altitudes of triangle *ABC*.

Multiplying the three equations above together we obtain Equation 11.2. Hence, altitudes *BG*, *AE* and *CF* are concurrent by Ceva.   □

It is rather remarkable that one can prove this Euclidean theorem using projective techniques. In the exercises, you may show similarly that medians and angle bisectors of a triangle are concurrent (Exercise 11.13).

Exercises

11.1   Find an automorphism of $\mathbf{P}^2(\mathbf{C})$ which is not a projective collineation. Justify.

11.2   Let **A** be an affine plane, and let **S** be its completion to a projective plane (Chapter 2). Show that a translation τ of **A** (Chapter 1) may be extended to an elation α of **S** by defining

$$\alpha(X) = \begin{cases} X & \text{if } X \in \ell_\infty \\ \tau(X) & \text{if } X \in \mathbf{A}. \end{cases}$$

11.3   Show that the set $\mathcal{E}(\ell)$ of elations with axis ℓ forms a group under composition of mappings $\pi \to \pi$.

11.4   Show that $\text{Tran}\,\mathbf{A} \cong \mathcal{E}(\ell)$ as groups, if ℓ is a line in a projective plane π, and **A** is the affine plane $\pi \smallsetminus \ell$ (cf. Exercise 2.2).

11.5   Find two elations, α and β, of the 7-point plane π_7 such that $\alpha \circ \beta$ is not an elation. *Hint:* Refer to Figure 4.2 and the proof of Proposition 4.18.

11.6   Let *H* be a subgroup of a group *G*. Show that for an arbitrary *g* in *G*
   (*a*) the set $gHg^{-1} = \{ghg^{-1} \mid h \in H\}$ is a subgroup of *G*;
   (*b*) the mapping $\phi \colon h \mapsto ghg^{-1}$ $(h \in H)$ is a homomorphism $\phi \colon H \to gHg^{-1}$;
   (*c*) ϕ is a bijection between *H* and gHg^{-1}.

11.7   Suppose π is a projective plane satisfying P5. Given $\phi \in \text{Aut}\,(\pi)$ and the group of elations $\mathcal{E}(\ell)$ with axis ℓ, show that the conjugate subgroup $\phi \mathcal{E}(\ell) \phi^{-1}$ is the group of elations $\mathcal{E}(\ell')$ with axis $\ell' = \phi(\ell)$.

11.8   Show in detail that

(a) if $G = H \rtimes K$, ϕ is an isomorphism $G \to G'$, $\phi(H) = H'$, $\phi(K) = K'$, then $G' = H' \rtimes K'$ (cf. Chapter 9, semi-direct product).

(b) $\mathcal{H}(\ell) = \mathcal{E}(\ell) \rtimes \mathcal{H}(\ell, O)$ (cf. Proposition 11.8).

11.9   Let $\pi = \mathbf{P}^2(\mathbf{F})$, $\mathbf{F}$ a field. Show that any homology $\delta \colon \pi \to \pi$ can be represented by a matrix similar (in the technical sense of linear algebra) to the matrix $\mathbf{C} = \begin{pmatrix} a & 0 & 0 \\ 0 & 1 & 0 \\ 0 & 0 & 1 \end{pmatrix}$, with $a \in \mathbf{F}$.

11.10   Prove that

(a) any invertible 3×3 matrix $\mathbf{Y}$ over a field $\mathbf{F}$ is a product of five special matrices: $\mathbf{Y} = \mathbf{C}_5\mathbf{C}_4\mathbf{A}_3\mathbf{A}_2\mathbf{A}_1$ where $\mathbf{C}_5$ and $\mathbf{C}_4$ are similar to $\mathbf{C}$, and $\mathbf{A}_3, \mathbf{A}_2, \mathbf{A}_1$ are similar to $\mathbf{A}$ in the proof of Theorem 11.12.

(b) Suppose a function $D \colon \mathrm{M}_3(\mathbf{F}) \to \mathbf{F}$ mapping the set of 3×3 matrices into $\mathbf{F}$ satisfies the two conditions

D1. For any two 3×3 matrices A, B,

$$D(AB) = D(A)D(B).$$

D2. For each $a \in \mathbf{F}$, let $\mathbf{C}(a) = \begin{pmatrix} a & 0 & 0 \\ 0 & 1 & 0 \\ 0 & 0 & 1 \end{pmatrix}$. Then $D(\mathbf{C}(a)) = a$.

Show that D is the determinant; i.e. $D(X) = \det(X)$ on every 3×3 matrix X.

11.11   Let $\pi = \mathbf{P}^2(\mathbf{F})$, $\mathbf{F}$ a field. Settle a question left hanging since Theorem 10.10 by defining cross ratio on any line m in π and extending Theorem 10.10 to arbitrary lines.

11.12   Let $\pi = \mathbf{P}^2(\mathbf{F})$, $\mathbf{F}$ a field of characteristic $\neq 2$. Let $A = (1, 0, 0)$, $B = (0, 1, 0)$, and $C = (0, 0, 1)$. Assume ℓ is a line with equation $ax_1 + cx_2 = 0$.

(a) Show that Theorem 11.13 holds true in this case, too. Why must $a \neq 0$ and $c \neq 0$?

(b) If one of a or $b = 0$, show that Theorem 11.13 is still true. Dispose of the outstanding cases in the proof of Theorem 11.13.

(c) What can go wrong in characteristic 2?

11.13   Let ABC be a triangle in the Euclidean plane.

(a) Prove that the medians of ABC are concurrent.

(b) Prove that the angle bisectors of ABC are concurrent.

11.14   Prove Menelaus' theorem:

Given a triangle ABC in the Euclidean plane and points U, V and W on BC, AC, and AB, respectively, then U, V and W are collinear iff

$$\frac{AW}{WB} \cdot \frac{BU}{UC} \cdot \frac{CV}{VA} = -1.$$

11.15   Suppose ABC is a triangle in the Euclidean plane, and X and Y are midpoints of the sides AB and BC, respectively. Show that $XY \parallel AC$.

11.16   Let $\mathbf{F}_q$ be a finite field with q elements. (It is shown in a full year algebra course that q is necessarily a power of a prime and $\mathbf{F}_q$ is unique up to isomorphism.) The finite projective plane $\mathbf{P}^2(\mathbf{F}_q)$ is called the projective plane of order q. Determine the number of projective collineations of the projective plane of order q.

11.17   A line in general position will intersect a complete quadrangle in six points called an *involutory hexad* or a *quadrangular set*. Define a projective invariant of ordered six-tuples on a line such that the six points form a quadrangular set if and only if your invariant is -1.

     Hint: Use Equation 11.1 and Figure 11.4 as a starting point.

Independent Studies in Projective Geometry

Projective geometry leads naturally to the study of algebraic geometry, non-Euclidean geometry or foundations of geometry. The authors would like to leave the student with an approach to each of these subjects. Information and references are provided in five appendices which we hope will inform the reader enough to profit from independent study. We also believe each of the five topics suitable for an accredited project.

Appendix A is about conics and Appendix B about Bezout's theorem. Both are topics at the beginning of a typical course in algebraic geometry, and might interest the reader in looking further into algebraic geometry (e.g. [Reid]). We have considered conics in the real projective plane in several places in the text and exercises; in Appendix A we then give an easily understood synthetic definition of conic that permits its study in any Pappian plane.

Having Bezout's theorem in one's geometry is the natural justification for assuming each pair of lines meet in one point. In a projective plane over an algebraically closed field, two algebraic curves of degree m and degree n meet in mn points (counting with multiplicity).

In Appendix C, we place a metric on the real projective plane and delve a little into the resulting elliptic geometry. Elliptic and spherical geometries are locally the same, globally different by the presence of Axiom P1 in the first. In both these geometries triangles have internal angles that sum to more than 180°. The hyperbolic geometry, where triangles have internal angles summing to less than 180°, has an incidence geometry represented by the Klein model in the projective plane — we will define this, too. We cover two cases of Bolyai's theorem, which says that the law of sines, suitably defined, holds for the Euclidean, spherical and hyperbolic geometries. A unified proof of Bolyai's sine law is given in [Hsiang].

In Appendix D we take up the evident question after a reading of Chapter 9: are there sensible coordinates for a projective plane without Desargues' theorem? It turns out that any projective plane π is coordinatized by a ternary ring R. The stronger the geometric axioms we put on π, the stronger the algebraic axioms we get on the generalized ring R. For example, suppose a projective plane π satisfies a modified version of P5:

> **P5*.** Let ABC and $A'B'C'$ be two triangles in π and O a point such that AA', BB', and CC' meet at O. Let $P = AB.A'B'$, $Q = AC.A'C'$ and $R = BC.B'C'$. Assume that O lies on PQ. Then R lies on PQ.

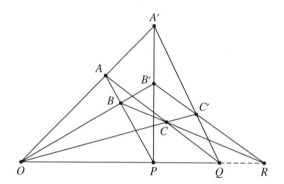

The configuration of P5*.

Such a π is called an *alternative plane*. Then π is isomorphic to a projective plane over an alternative division ring, like the Cayley-Graves octonions. The corresponding affine axiom is in fact A4, and the algebraic conditions we get on a ternary ring are presented in Appendix D.

In Appendix E, we show that projective geometry may be viewed as a study of a particular kind of lattice theory — the lattice of subspaces of a vector space. There is also a principle of duality for this wider class of lattices; indeed, it is mostly through these means of generalizing the original principle of duality in projective geometry that duality has come to be a notion pervading modern mathematics. In Chapter 8, we showed that an automorphism of a projective plane with P5 is a semilinear transformation: in Appendix E, we present a vast generalization of this fundamental theorem to isomorphisms of projective geometries.

There are other projects suggested in Exercises 8.13 and 9.4. The project in Exercise 8.13 is to convert the fundamental theorem of 2-dimensional projective geometry to n-dimensional projective geometry. This project is closely related to the one in Appendix E. The project in Exercise 9.4 involves studying some more group theory and seeing why the 168-element automorphism group of the seven point plane is a simple group.

Appendix A

Conics

In this appendix we will give five different definitions of a conic. Each will be a different aspect of the same figures we are all familiar with — the circle, ellipse, parabola and hyperbola. Several of the definitions will lead to generalizations of the ordinary conic to finite geometry, complex geometry, or other specializations of a field **F**. The challenge for you is to show the equivalence of some or all of the definitions, over the course of several weeks or months. We include a sketch of the proofs of several implications and mention several important projective theorems you should obtain along the way. Finally, a set of exercises at the end will be helpful in illuminating the text.

A.1 Conics

In the Euclidean plane, a first notion is a circle since it is the locus of all points equidistant from a point. As a second step, conic is naturally defined as a figure obtained from a circle by applying a series of finitely many central or parallel projections in Euclidean space (and ending up in the plane of the original circle):

A.1. Desargues' definition A *conic* is the locus of points in a plane obtained from a circle after a finite number of projections in space.

According to Exercise 5.8 one can deduce from this definition a sensible cross ratio of four coconic points, say points A, B, C, D on conic Γ. $R_x(A,B;C,D)$ is defined to be $R_x(AP,BP;CP,DP)$ for any fifth point P on Γ: Exercise 5.8 shows the choice of P in the cross ratio of four lines concurrent at P to be irrelevant. Thus

$$R_x(AX,BX;CX,DX) = \text{constant} \qquad \text{if } X \in \Gamma \smallsetminus \{A,B,C,D\}.$$

Indeed the converse is not hard to prove, either (cf. Exercise A.2).

A.2. Chasles's definition Suppose A, B, C, D are points in a plane such that no three are collinear. A *conic* is the locus of points A, B, C, D and fifth varible point X such that

$$R_x(AX, BX; CX, DX) = k \qquad (k \neq 0, 1).$$

Given two points A, B on a conic Γ, a projectivity between pencils at A and B, $\tau \colon [A] \barwedge [B]$, may be defined as follows: given $X \in \Gamma$,

$$\tau \colon AX \mapsto BX.$$

That leaves only the small detail of the tangent lines to Γ at A and B (why?). τ sends AB into the tangent line b at B, and sends the tangent line a at A into AB. Now why is the one-to-one correspondence τ a projectivity? Basically because it is a cross ratio preserving transformation of pencils: just apply Definition A.2 to four coconic ponts! It is not hard to prove that Definition A.3 below is equivalent to Definition A.2 in the Euclidean plane.

A.3. Steiner's definition Let τ be a projectivity between pencils of lines centered at points A and B. Suppose $A \neq B$ and τ is not a perspectivity. A *conic* Γ is the locus of points $\ell . \ell'$ where $\tau \colon \ell \barwedge \ell'$.

Definition A.3 fits very well into our synthetic development of projective geometry. Thus we have here a natural definition of conic in any projective plane. However, it appears that A and B play some special role on Γ, whereas the truth is that for any two points C and D on Γ in a Pappian plane there is a projectivity $\sigma \colon [C] \barwedge [D]$ such that $\Gamma = \{X \mid \sigma \colon CX \barwedge DX\}$: we might suggestively summarize this by writing $\Gamma = \Gamma(A, B; \tau) = \Gamma(C, D; \sigma)$. Since three points and their values determine a projectivity in a Pappian plane, we might write this as $\Gamma = \Gamma(A, B; C, D, E)$, where A, B, C, D, E are five distinct points and τ is determined from $AC, AD, AE \barwedge BC, BD, BE$.

At this point, you should prove

A.4. Pascal's theorem *Let Γ be a conic, and A, B, C, D, E points on Γ. Given a sixth point F in the projective plane, consider the hexagon ABCDEF. Suppose no three vertices are collinear. Then $F \in \Gamma$ if and only if AB.DE, BC.EF and CD.AF are collinear.*

You may essentially copy the proof of Pappus' theorem in Chapter 6 to obtain the proof of the "only if" part, if you assume $\Gamma = \Gamma(A, C; \tau)$.

Now you should attempt to show that you may cyclically permute the points in $\Gamma = \Gamma(A, B; C, D, E)$; thus showing Γ not to be dependent on A and B. Help may be found in [Seidenberg].

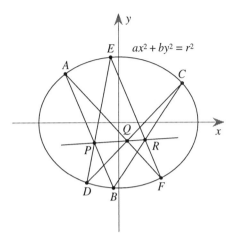

Figure A.1. Pascal's theorem.

Definition A.3 also suffers from the drawback of not being visibly self-dual. If you dualize Definition A.3, you would obtain something you could call a line-conic, the envelope of lines through projectively related points on ℓ and ℓ'. You have already seen an instance of the dual of Definition A.3 in Exercise 5.5. If you define a point-conic to be a conic with its tangent lines (their definition?), and include the contact points (their definition?) in line-conic, it is possible to prove that point-conics are line-conics, and line-conics are point-conics. This is what is meant when we say conic is a self-dual concept. So dualizing Pascal's theorem gives a theorem stating roughly:

A.5. Brianchon's theorem *Opposite vertices of a hexagon circumscribed to a conic join in three concurrent lines.*

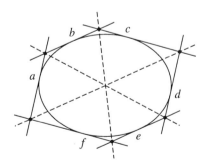

Figure A.2. Brianchon's theorem

This theorem was discovered around 1806 while Pascal's theorem predated it by as much as 167 years. It was in fact Brianchon's theorem that gelled the principle of duality in the minds of Poncelet, Gergonne and others: see [Bell].

Let us cast about for a self-dual definition of conic. In Exercise A.3 you associate to a point P, away from a conic Γ, the line p obtained as the locus of the variable point Z on a secant ℓ such that $H(X,Y;Z,P)$, where $\{X,Y\} = \Gamma \cap \ell$.

We call the line p the *polar* of P, and P the *pole* of p. Points on Γ receive their tangent lines as poles.

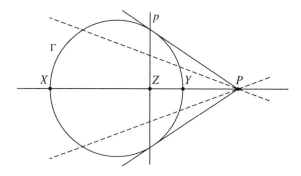

Figure A.3. Constructing the polar of a point with respect to a conic.

The assignment of pole to polar, polar to pole, with respect to a conic, is an example of a *polarity*. A polarity is a one-to-one correspondence between points and lines of a projective plane, respecting incidence: thus a polarity sends ranges into pencils, pencils into ranges, and complete quadrangles into complete quadrilaterals, etc. Moreover, one assumes of a polarity that it sends each range of points into a pencil of lines *projectively*. One generalizes terminology to say that a polarity sends A to its polar a, and sends a line b to its pole B.

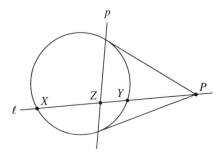

Figure A.4. The pole p of P with respect to a conic Γ.

A polarity has the property that if it sends a into A, then it sends A into a. We say A and B are *conjugate points*, and a, b *conjugate lines* of a polarity T if $B \in a$

and $A \in b$ (and by convention $T(B) = b$, $T(A) = a$). So a point X is *self-conjugate* if $X \in x$. A polarity is termed *hyperbolic* if it has at least one self-conjugate point. Then one can show that every line but one has exactly two self-conjugate points. Self-conjugate lines are similarly defined, and their theory is developed dually.

A.6. von Staudt's definition A conic is the locus of self-conjugate points of a hyperbolic polarity.

If we include the envelope of self-conjugate lines to this definition — as we include the lines to the vertices of a projectively defined triangle — we will obtain a self-dual definition of conic. Showing Definitions A.3 and A.6 equivalent will then complete the program of understanding self-duality of conic within our synthetic development, as well as add much to your understanding of conics. You will want to prove some basic theorems as possible stepping stones to proving Definition A.3 equivalent to A.6; we mention them below (and do Exercises A.4 and A.5 as well). Help may be found in [Coxeter].

A.7. von Staudt's theorem *Given a polarity, there exists a self-polar triangle and a unique self-polar pentagon. (An odd polygon is said to be self-polar if each vertex is transformed into the opposite side.)*

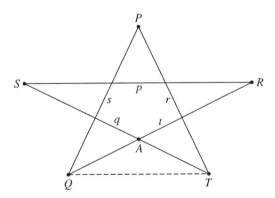

Figure A.5. The polar diagrammed is equally well described by $(AQT)(Pp)$ for the self-polar triangle AQT with extra value $P \mapsto p$, or by the self-polar pentagon $PTSRQ$.

A.8. Chasles' theorem *Given a polarity, a triangle and its polar triangle, if distinct, are in perspective.*

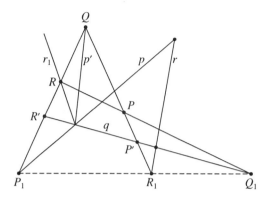

Figure A.6. Given triangle PQR and polar triangle pqr, show that P_1, R_1 and Q_1 are collinear.

A.9. Desargues' involution theorem *Suppose PQRS is a complete quadrangle in a Pappian plane and ℓ a line not passing through either P, Q, R or S. Each conic Γ through P,Q,R,S that meets ℓ in a pair of points X,Y does so in pairs of an involution τ.*

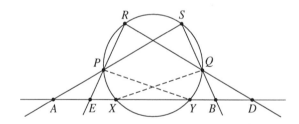

Figure A.7. Pairs of points belonging to an involution on a line arise from a "pencil" of conics through four points.

In fact, using Exercises 5.8 and 5.18 one proves that

$$AEXY \barwedge DBYX.$$

This must be an involution by Exercise 6.7.

We now ask, what is the situation in coordinates over a field **F**? You should show that a polarity of $\mathbf{P}^2(\mathbf{F})$ is represented by a symmetric, invertible matrix. Then you can establish from Definition A.6 the equivalence of the next definition with the others.

A.10. Definition A *conic* is the locus of points $X = (x_1, x_2, x_3)$ such that

$$XCX^{\mathrm{T}} = 0,$$

$$c_{11}x_1^2 + c_{22}x_2^2 + c_{33}x_3^2 + 2c_{23}x_2x_3 + 2c_{31}x_3x_1 + 2c_{12}x_1x_2 = 0,$$

for some symmetric matrix $C = (c_{ij})$ with nonzero determinant.

In the real projective plane you should show that every conic transforms under an automorphism to

$$x_1x_3 = x_2^2.$$

See [Reid] for more on the equivalence of conics, and quadratic forms, their classification, and why a conic is isomorphic to a projective line.

Exercises

A.1   Suppose that four concurrent lines $\ell_1, \ell_2, \ell_3, \ell_4$ in the coordinatized Euclidean plane have slopes m_1, m_2, m_3, m_4, respectively. Show that the cross ratio satisfies

$$R_x(\ell_1, \ell_2; \ell_3, \ell_4) = \frac{m_1 - m_3}{m_2 - m_3} \Big/ \frac{m_1 - m_4}{m_2 - m_4}.$$

   Hint: Use Exercise 5.7.

A.2   Consider the four points $A = (1,0)$, $B = (-1,0)$, $C = (0,1)$, $D = (0,-1)$ in the Euclidean plane, and a variable point $P = (x,y)$ such that

$$R_x(PA, PB; PC, PD) = -1.$$

   Compute the locus of points satisfying this equation.

A.3   Refer to Exercise 5.11 and Figure 5.14. Let ℓ be a secant line through P and intersecting circle C at Z and W.
   (*a*) Prove that if $Y = A_1A_2.\ell$, then $H(Z,W;Y,P)$.
   (*b*) Prove that $\angle XBA_1$ is a right angle.

A.4   Let A, B, C be distinct points on a conic Γ. Let a, b, c be tangents of Γ at A, B, C, respectively. Show that $AB.c$, $AC.b$ and $BC.a$ are collinear.
   Hint: Apply Chasles' theorem.

A.5   Given a polarity, a line ℓ, and $A \in \ell$, associate to A the point $A' = a.\ell$. Show that $A \mapsto A'$ defines an involution of ℓ (cf. Exercise 6.7).

A.6   Let π be a finite projective plane of order q. Show that the number of conics is $q^5 - q^2$.

Appendix B

Algebraic Curves and Bezout's Theorem

B.1 Algebraic Curves

An *algebraic curve* in the affine plane $\mathbf{A}^2(\mathbf{R})$ is the locus of points (x,y) satisfying a polynomial equation in two variables and real coefficients,

$$f(x,y) = 0.$$

Algebraic curves we have seen so far include points, $(x-a)^2 + (y-b)^2 = 0$; lines, $ax + by + c = 0$; triangles, $(ax + by + c)(a'x + b'y + c')(a''x + b''y + c'') = 0$; and conic sections, $ax^2 + by^2 + cxy + dx + ey + f = 0$ (with certain *degenerate conics* occuring for some choices of $a, b, \ldots, f$).

Algebraic curves we have not yet seen include *elliptic curves*, a cubic curve of the form

$$y^2 = f(x) = x^3 + ax^2 + bx + c,$$

where the roots of $f(x)$ are distinct (complex) numbers. Fermat's last theorem, a long-standing conjecture about the triviality of the set of rational solutions to the equation

$$x^n + y^n = 1 \qquad (n \geq 3),$$

has recently been reduced to a theorem about the group of rational points on an elliptic curve with rational coefficients: for further information, see [Silverman-Tate] and [Rubin-Silverberg].

Naturally, there are many other algebraic curves $f(x,y) = 0$ of higher degree, the *degree* being defined to be the highest sum of powers $i + j$ present among the monomials $x^i y^j$ occurring in $f(x,y)$.

159

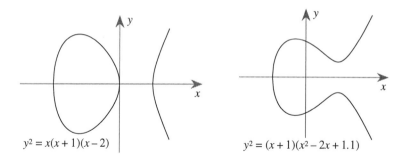

$$y^2 = x(x+1)(x-2) \qquad\qquad y^2 = (x+1)(x^2-2x+1.1)$$

Figure B.1. Two elliptic curves.

In the projective plane $\mathbf{P}^2(\mathbf{R})$ there is the corresponding notion of algebraic curve. These are also the zeros of polynomials like in the affine case, just with the difference that the polynomials must be homogeneous polynomials of three variables in order to assume consistency with the definition of a point as a line in $\mathbf{R}^3$ through $(0,0,0)$. A polynomial $F(X,Y,Z)$ is called a *homogeneous polynomial* of *degree d* if and only if F satisfies

$$F(tX,tY,tZ) = t^d F(X,Y,Z) \qquad (\forall t).$$

This implies that $F(X,Y,Z)$ must be a sum of monomials $X^i Y^j Z^k$ where $i+j+k = d$ is fixed. Then define a *projective algebraic curve* of *degree d* as the locus of points with homogeneous coordinates (X,Y,Z) such that

$$F(X,Y,Z) = 0.$$

An algebraic curve in $\mathbf{P}^2(\mathbf{R})$, say $F(X,Y,Z) = 0$, corresponds to an affine algebraic curve, $f(x,y) = 0$, by *dehomogenization* (with respect to a variable, say Z), a process you will recognize from switching a line to its affine coordinate equation. One simply sets $Z = 1$, letting $f(x,y) = F(x,y,1)$. For example, the conic $X^2 + Y^2 - Z^2 = 0$ takes either the form of a circle, $x^2 + y^2 = 1$, or of a hyperbola, $x^2 - z^2 = -1$, if we dehomogenize with respect to Y.

It can happen that much information is lost in dehomogenization. In an extreme example, $F(X,Y,Z) = Z = 0$ should dehomogenize to $1 = 0$, which should be interpreted as the line at infinity of the affine plane $Z \neq 0$. So we need to think of the affine curve corresponding to a projective curve as the curve $F(x,y,1) = 0$ together with the ideal points $(X,Y,0)$ satisfying $F(X,Y,0) = 0$. For example, $F(X,Y,Z) = X^2 - Y^2 + Z^2 = 0$ dehomogenizes to the affine algebraic curve $f(x,y) = x^2 - y^2 + 1$ together with the ideal points $(1,\pm1,0)$.

To an affine algebraic curve

$$f(x,y) = \sum a_{ij} x^i y^j = 0$$

of degree d, we make correspond the projective curve

$$F(X,Y,Z) = \sum a_{ij}X^iY^jZ^{d-i-j} = 0.$$

This correspondence is called *homogenization* and is clearly inverse to dehomogenization in the sense that a one-to-one correspondence is set up between affine and projective algebraic curves that do not contain the line $Z = 0$.

It may be of great value to transform one algebraic curve of degree d to another by a projective collineation. For example, the curve

$$C: X^2 + 2Y^2 + 3Z^2 + 2XY + 2XZ + 4YZ = 0$$

transforms to $(X')^2 + (Y')^2 + (Z')^2 = 0$, which clearly has no solutions: since projective collineations — in fact we used $X' = X + Y + Z$, $Y' = Y + Z$, $Z' = Z$ — are invertible, we conclude there is no solution to the first equation, either.

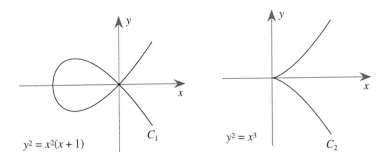

Figure B.2. Singularities at $(0,0)$.

The curves above, $C_1: y^2 - x^3 - x^2 = 0$ and $C_2: y^2 - x^3 = 0$, as well as their homogenizations $\tilde{C}_1: Y^2Z - X^3 - X^2Z = 0$ and $\tilde{C}_2: Y^2Z - X^3 = 0$, are said to have *singularities* at the point $(0,0)$ or $(0,0,1)$, because all their partial derivatives vanish at these points. For example,

$$\frac{\partial}{\partial Y}(Y^2Z - X^3 - X^2)\Big|_{(0,0,1)} = 2YZ|_{(0,0,1)} = 0.$$

An algebraic curve with no singularities is said to be *smooth*.

Algebraic curves may be defined in an entirely similar way over a general field. Smoothness carries over, too, in spite of the seeming use of limit and ε-δ arguments with a notion of distance implicit. Since we deal only with polynomials, we can define partial differentiation formally by $\frac{\partial}{\partial X}X^nY^iZ^j = nX^{n-1}Y^iZ^j$, etc.

B.2 Intersection Theory of Algebraic Curves

We can now investigate how many points of intersection there are between two projective algebraic curves of degree m and n. If $m = n = 1$ and the lines are different, we know there is only one answer (unlike the affine case): one point.

If $m = 1$ and $n = 2$, we can run into the following difficulty: if $C_1 : X - Y = 0$ and $C_2 : X^2 - Y^2 = 0$, then $C_1 \cap C_2$ contains an infinite number of points — in fact, the whole curve C_1. Indeed C_2 is the union of two lines, one of them being C_1. The way to circumvent such a nuisance is to remark that, like unique factorization of integers into primes, polynomials may be factored into irreducible polynomials (of degree 1 or more, and homogeneous in the projective case). For example, $X^2 - Y^2 = (X + Y)(X - Y)$. We now insist that we look only at pairs of curves with no common component, i.e., no one polynomial occurs in both factorizations into irreducible polynomials.

Now another obvious occurrence when $m = 2$ and $n = 1$ is that the line and conic might miss one another entirely. For example $C_1 : X^2 + Y^2 - Z^2 = 0$ and $C_2 : X - 3Z = 0$. Although $C_1 \cap C_2$ contain no real points, if we allow complex solutions, i.e., we consider C_1 and C_2 as curves in $\mathbf{P}^2(\mathbf{C})$, then we get two solutions, viz. $(3, \pm i\sqrt{8}, 1)$.

Yet another sort of example is the following: $C_1 : Y - Z = 0$ and $C_2 : X^2 + Y^2 - Z^2 = 0$. This is a circle and one of its tangent lines when we homogenize with respect to Z. It is certain that $C_1 \cap C_2 = \{(0, 1, 1)\}$, but perhaps $(0, 1, 1)$ should count twice! After all, if we substitute $Y - Z = 0$ in $X^2 + (Y - Z)(Y + Z) = 0$, we get $X^2 = 0$, and we would say $X = 0$ is a *root of multiplicity* 2 (i.e., also a root of the derived polynomial). Multiplicity in intersection theory is captured by the notion of intersection multiplicity at a common point P, denoted by $I(C_1 \cap C_2, P)$. Its definition is technical, involving rings called local rings, and postponed until the exercises where we look only at the affine plane: a highly recommended reference for beginnners is [Silverman-Tate, Appendix A].

Let it suffice to say that $I(C_1 \cap C_2, P) = 1$ if C_1 and C_2 intersect *transversally* at P. In affine terms, this means that P is a nonsingular point for both curves and their tangents span the plane as in Figure B.3.

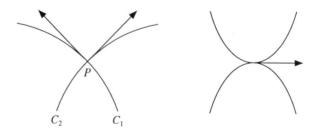

Figure B.3. Left: Transversal intersection. Right: nontransversal intersection.

We are now ready to state the definitive theorem in the subject, Bezout's theorem for real algebraic curves viewed in $\mathbf{P}^2(\mathbf{C})$.

B.1. Bezout's theorem *Let C_1 and C_2 be projective algebraic curves with no common components. Then*

$$\sum_{P \in C_1 \cap C_2} I(C_1 \cap C_2, P) = (\deg C_1)(\deg C_2).$$

In particular, if C_1 and C_2 are smooth with only transversal intersections, then $\#(C_1 \cap C_2) = (\deg C_1)(\deg C_2)$. *In all other cases,*

$$\#(C_1 \cap C_2) \leq (\deg C_1)(\deg C_2).$$

A proof of Bezout's theorem has been broken into a series of easy exercises in [Silverman-Tate]: we recommend the student focus in this project on the meaning of Bezout's theorem, its proof, its applications, and its history. The student should see [Reid] for a special proof of Bezout's theorem in case $\deg C_1 = 2 = \deg C_2$.

Bezout's theorem is very powerful. For example, let us apply it to two curves C_1 and C_2 both of degree 2. If we moreover suppose that both are conics and have five points in common, no three of which are collinear, then Bezout's theorem tells us that $C_1 = C_2$. Moreover, it is possible to prove the following theorem when $d_1 = d_2 = 3$ using an argument that identifies the set of cubic curves with $\mathbf{P}^9(\mathbf{R})$ (see [Silverman-Tate]).

B.2. Bacharach-Cayley theorem *Let C_1 and C_2 be projective algebraic curves of degrees d_1 and d_2, without common components. Suppose that C_1 and C_2 intersect in $d_1 d_2$ points. Let $\mathcal{D}$ be a projective algebraic curve of degree $d_1 + d_2 - 3$. If $\mathcal{D}$ passes through $d_1 d_2 - 1$ points of $C_1 \cap C_2$, then it passes also through the remaining point of $C_1 \cap C_2$.*

Let us apply the Bacharach-Cayley theorem and Bezout's theorem to prove (one half of) Pascal's theorem. The following proof would in principle be valid in $\mathbf{P}^2(\mathbf{F})$ for any field $\mathbf{F}$ (of characteristic $\neq 2$) since we can replace $\mathbf{R}$ and $\mathbf{C}$ with $\mathbf{F}$ and its algebraic closure[1] in the statement and proof of Bezout's theorem and the Bacharach-Cayley theorem.

B.3. Pascal's theorem *Let C be a smooth conic and A, B, C, D, E, F six distinct points on C. Let $Q = AB.DE$, $R = BC.EF$ and $S = CD.AF$. Then Q, R and S are collinear.*

---

[1] See [Kaplansky, p. 74–76] for a proof that any field $\mathbf{F}$ has an algebraic closure in which degree n polynomials all factor into linear factors.

Demonstration Consider the cubic curves (not irreducible!) $C_1 = AB \cup CD \cup EF$ and $C_2 = BC \cup DE \cup AF$. All nine points $A, B, C, D, E, F, Q, R, S$ lie on C_1 and C_2. Let C_3 be the cubic curve defined by

$$C_3 = C \cup QR.$$

Now C_3 contains eight of the points above; viz., A, B, C, D, E, F, Q and R. By the Bacharach-Cayley theorem, C_3 contains S, too. Where on C_3 is S? If $S \in C$, the line AF intersects C in three distinct points in contradiction of Bezout's theorem. We conclude that $S \in QR$, demonstrating collinearity.   □

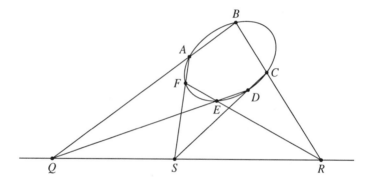

Figure B.4. Pascal's theorem.

Exercises

B.1   Find the well-known formula for a solution x of $Ax^2 + Bx + C = 0$ by applying a projective transformation (projectivity) of the projective line.

  Hint: Multiply by $4A$ and complete the square.

B.2   Let R be a ring. An *ideal* I in R is a subset of R such that

  I1. $x - y \in I$ for all $x, y \in I$.
  I2. $xy \in I$ and $yx \in I$ for all $x \in R$, $y \in I$.

Check that

  (a) I is a ring in itself, though possibly without 1.
  (b) I is a normal subgroup in R under $+$.
  (c) The set of cosets R/I gets a well-defined multiplication as follows:

$$(a + I)(b + I) = ab + I.$$

  (d) R/I is a ring.

(e) If $f: R \to S$ is a homomorphism of rings, which is surjective, then $\mathrm{Ker}(f)$ is an ideal in R.

(f) If f is surjective, then $R/\mathrm{Ker}(f) \cong S$.

B.3   Let R be a commutative ring. A subset T of R is said to be *multiplicative* if $1 \in T$, $0 \notin T$ and $ab \in T$ for all $a, b \in T$. An ideal P in R is said to be a *prime ideal* if $ab \in P$ implies $a \in P$ or $b \in P$. Show that

(a) P is a prime ideal if and only if $R \smallsetminus P$ is a multiplicative set.

(b) If R is the ring of integers, ideals are of the form $(n) = \{nq \mid q \in \mathbf{Z}\}$, and prime ideals of the form (p), p a prime number.

B.4   This exercise generalizes an earlier exercise on forming the field of fractions. Given a multiplicative subset T of a ring R, we form the *ring of fractions* of R by T as follows. In $R \times T$ define a relation by

$$(r,t) \sim (r',t')$$

if there exists an element t_1 in T such that

$$t_1(t'r - tr') = 0.$$

(a) Check that $\sim$ is an equivalence relation.

(b) Denote the equivalence class of (r,t) by $\frac{r}{t}$, and the set of these by R_T. Define a multiplication in R_T by

$$\frac{r}{t} \cdot \frac{r'}{t'} = \frac{rr'}{tt'}.$$

Check that multiplication is independent of representative chosen for the equivalence class.

(c) Define addition in R_T by the rule

$$\frac{r}{t} + \frac{r'}{t'} = \frac{t'r + tr'}{tt'}.$$

Check that addition is well-defined and R_T is a ring.

(d) Check that the mapping $f: R \to R_T, r \mapsto \frac{r}{1}$ is a homomorphism.

(e) Let S be another multiplicative subset of R and $S \subseteq T$. Define a natural mapping $R_S \to R_T$.

B.5   An ideal M in a commutative ring R is said to be *maximal* if $K = M$ or $K = R$ for every ideal K such that $M \subset K \subset R$.

(a) Prove that if M is a maximal ideal, then M is a prime ideal.

(b) R/M is a field.

(c) P is a prime ideal if and only if R/P has no zero divisors.

(d) One can form the ring of fractions of R by $R \smallsetminus M$, which is traditionally denoted by R_M.

(e) If $R = \mathbf{F}[X,Y], P = (X) = \{Xf \mid f \in R\}, M = (X,Y) = \{Xf_1 + Yf_2 \mid f_1, f_2 \in R\}$, then P is prime but not maximal and M is maximal.

Let $C_1 : f_1(x,y) = 0$ and $C_2 : f_2(x,y) = 0$ be two affine algebraic curves with no common factor. So $f_1, f_2 \in \mathbf{F}[x,y]$, the polynomial ring over an algebraically closed field $\mathbf{F}$ in two indeterminates. Suppose $P \in C_1 \cap C_2$. Let R be the commutative ring $\mathbf{F}[X,Y]$, which is in fact an integral domain. Form its field of fractions K, denoted by $\mathbf{F}(X,Y)$ and called the *field of rational functions* of X and Y. In R there is a maximal ideal $M(P) = \{f(x,y) \mid f(P) = 0\}$: check this. Then the ring of fractions $R_{M(P)} = O_P$ is called the *local ring* of P.

Check that O_P is a subring of K via the map you defined in B.4e. Check that the map

$$\frac{f(x,y)}{g(x,y)} \mapsto \frac{f(P)}{g(P)}$$

defines a homomorphism of O_P onto $\mathbf{F}$ with kernel

$$M_P = \left\{ \frac{f}{g} \in O_P \,\middle|\, \frac{f(P)}{g(P)} = 0 \right\}.$$

Now let (f_1, f_2) be the ideal in O_P generated by $f_1/1$ and $f_2/1$ in O_P: i.e.

$$(f_1, f_2) = \left\{ \frac{f_1 f}{g} + \frac{f_2 f'}{g'} \,\middle|\, f, f' \in R,\ g, g' \in R \setminus M(P) \right\}.$$

The *intersection multiplicity* of C_1 and C_2 at P is

$$\mathrm{I}(C_1 \cap C_2, P) = \dim \left(\frac{O_P}{(f_1, f_2)} \right)$$

where the right hand side is just the dimension of a vector space, since the field $\mathbf{F}$ acts like scalars.

B.6   Compute $\mathrm{I}(C_1 \cap C_2, P) = 1$ where $C_1 : x = 0$, $C_2 : y = 0$ and $P = (0,0)$.

B.7   Let $C_1 : y = x^2$ and $C_2 : y = x^3$. Homogenize to projective curves, also denoted by C_1 and C_2. Show by any means that $\sum_{P \in C_1 \cap C_2} \mathrm{I}(C_1 \cap C_2, P) = 6$.

Appendix C

Elliptic Geometry

C.1 The Laws of Sines and of Cosines

Euclidean geometry has taught us that planar geometry is especially fertile in the presence of a notion of distance between points and an angle between lines. You will recall some of the important theorems in Euclidean geometry: the law of cosines, the law of sines, and the theorem which states that the sum of interior angles of a triangle is π radians. For the convenience of the reader we state these theorems in Equations C.1 and C.2 below:

Law of cosines: $a^2 = b^2 + c^2 - 2bc\cos A$ (C.1)

Law of sines: $\dfrac{\sin A}{a} = \dfrac{\sin B}{b} = \dfrac{\sin C}{c}$ (C.2)

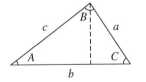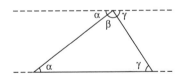

Figure C.1. Left: Angles and lengths of general triangle. Right: $\alpha + \beta + \gamma = \pi$.

The real projective plane has a respectable distance geometry on it as well: it is essentially the spherical geometry of the everyday experience of airline pilots! We use the sphere model of the real projective plane, denoted by $\mathbf{P}^2$ in this appendix. Let S^2 be the unit sphere in $\mathbf{R}^3$ centered at the origin O. A point P in $\mathbf{P}^2$ stands for a set of two antipodal points, $\pm P$, in S^2. A line ℓ in $\mathbf{P}^2$ is a great circle on S^2. A point $P \in S^2$ also determines a unit vector from O to P, which we also denote by P.

The distance between two points P and Q in $\mathbf{P}^2$ is defined to be the acute angle in radians between lines OP and OQ:

$$\mathrm{dist}(P,Q) = \arccos |P \cdot Q|.$$

Angles between lines is convertible to distance between points by a choice of polarity (or what is equivalent, a conic or nondegenerate quadratic form). Now the coordinates suggest a natural choice for the polarity: send point

$$(A, B, C) \mapsto \text{line } \ell \colon Ax + By + Cz = 0.$$

Inversely, a line or great circle on S^2 is sent to its "north-south pole:" the great circle through P and Q is sent to $P \times Q$. Then the angles between two lines a and b is the distance $\mathrm{dist}(A, B)$ between its poles. As a consequence the polar of a triangle XYZ with lengths a, b, c and angle measures α, β, γ is the triangle xyz with lengths α, β, γ and angle measures a, b, c.

We are going to derive the law of cosines and the law of sines for elliptic geometry and indicate in the exercises what the sum of internal angles is in a triangle. We will note the law of sines to be one instance of the great theorem of J. Bolyai (1802–1860). In his work on absolute geometry J. Bolyai states the following theorem for a general triangle in the spherical, Euclidean or hyperbolic planes (notation as in Figure C.1, left):

$$\frac{\sin A}{\odot a} = \frac{\sin B}{\odot b} = \frac{\sin C}{\odot c},$$

where $\odot r$ denotes the arclength of a circle of radius r in each geometry ($2\pi \sin r$, $2\pi r$ and $2\pi \sinh r$, respectively).

We suggest the following project: research the statement, history and proof of J. Bolyai's sine law, and then look at modern treatments such as the unified proof in [Hsiang]. We recommend [Ryan] for a modern textbook treatment of the basics of the Euclidean, spherical, elliptic and hyperbolic planes.

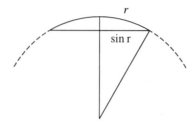

Figure C.2. Arclength $\odot r = 2\pi \sin r$ on sphere.

C.1. Law of cosines *For an elliptic triangle ABC with sides of length a, b, c and angle measures α, β, γ the following equation holds:*

$$\cos a = \cos b \cos c + \sin b \sin c \cos \alpha \qquad (C.3)$$

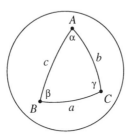

Figure C.3. A geodesic triangle on the sphere.

Proof Do Exercises C.1, C.2 and C.3 in order to establish the following advanced identity from vector analysis:

$$(C \times A) \cdot (A \times B) = (C \cdot A)(A \cdot B) - (C \cdot B)(A \cdot A). \qquad (C.4)$$

The left-hand side of (C.4) simplifies in three steps:

$$|C \times A|^2 = -(C \times A) \cdot (A \times C) = -(C \cdot A)^2 + (C \cdot C)(A \cdot A)$$
$$= -\cos^2 b + 1 = \sin^2 b,$$

so

$$|A \times B|^2 = \sin^2 c.$$

Hence

$$(C \times A) \cdot (A \times B) = -\sin b \sin c \cos \alpha.$$

The right-hand side is just

$$(C \cdot A)(A \cdot B) - (C \cdot B)(A \cdot A) = \cos b \cos c - \cos a,$$

whence (C.3) follows from (C.4).   □

The law of cosines in Euclidean geometry may be recovered from Equation C.3 by replacing $\cos x$ with $1 - x^2/2$ and $\sin y$ with y, their second degree Taylor polynomials.

C.2. Corollary *For the elliptic triangle above with $\gamma = \pi/2$, the following equations hold:*

$$\cos c = \cos a \cos b \qquad \text{(C.5)}$$
$$\cos \alpha = \sin \beta \cos a \qquad \text{(C.6)}$$
$$\sin b = \sin c \sin \beta \qquad \text{(C.7)}$$

Proof Note that the law of cosines gives three equations for one triangle; one of which is

$$\cos c = \cos a \cos b + \sin a \sin b \cos \gamma.$$

But $\cos \gamma = 0$ when $\gamma = \pi/2$, whence (C.5).

The law of cosines applied to the polar triangle gives

$$\cos \alpha = \cos \beta \cos \gamma + \sin \beta \sin \gamma \cos a.$$

But $\sin \gamma = 1$ when $\gamma = \pi/2$, whence (C.6).

Now apply (C.5) to (C.3), using $\cos^2 b = 1 - \sin^2 b$: we get

$$\cos a \sin b = \sin c \cos \alpha. \qquad \text{(C.8)}$$

Apply (C.6) to (C.8) and cancel to get (C.7). □

C.3. Law of sines *For an elliptic triangle (Figure C.4),*

$$\frac{\sin \alpha}{\sin a} = \frac{\sin \beta}{\sin b} = \frac{\sin \gamma}{\sin c}.$$

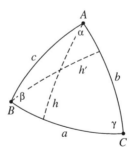

Figure C.4. Elliptic triangle with altitudes.

Proof Let h be the length of an altitude from A to BC. By (C.7) of the corollary we have

$$\sin \gamma \sin b = \sin h = \sin \beta \sin c.$$

Hence,

$$\frac{\sin \gamma}{\sin c} = \frac{\sin \beta}{\sin b}.$$

By dropping an altitude from B of length h' as in Figure C.4, you similarly prove that

$$\frac{\sin \gamma}{\sin c} = \frac{\sin \alpha}{\sin a}.$$

$\square$

C.2   The Incidence Geometry of the Hyperbolic Plane

Already in Chapter 2, we saw the incidence geometry of the Euclidean plane embedded in the real projective plane as a subgeometry (cf. Exercise 2.2). In the section just completed, we noted that spherical and projective geometry are locally identical — even in their metrical aspect. In this section, we will give a simple, brief description of the incidence geometry of the hyperbolic plane as a subgeometry of the real projective plane.[1]

Consider the subgeometry of points and lines in the "inside" of a conic in $\mathbf{P}^2$, i.e., the component of the complement of the conic that does not contain any full line. One concrete example of such is the set of points $P = (x, y, z) \in \mathbf{P}^2$ such that $x^2 + y^2 < 1$ and $z = 1$. Lines in this geometry $\{(x, y, 1) \mid x^2 + y^2 < 1\}$ are simply the line segments satisfying linear equations $ax + by + c = 0$. This is the *Beltrami-Klein model* of the hyperbolic plane H^2.

It is clear that two points in H^2 determine a unique line through them. Unlike Euclidean and projective geometries, there is not 0 or 1 lines through a point P and parallel to (disjoint with) a line ℓ such that $P \notin \ell$: there are infinitely many lines! We call all such lines $m \ni P$ such that $m \cap \ell = \emptyset$ *ultraparallel* to ℓ, except two, which are said to be *parallel* to ℓ: namely, the lines n_1 and n_2 that intersect ℓ on the circle boundary.

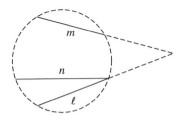

Figure C.5. Klein model: n and ℓ parallel lines; m and ℓ ultraparallel.

---

[1] It was these observations and others in which Arthur Cayley (1821–1895) obtained various metrics from the cross ratio that led him to pronounce with characteristic enthusiasm metric geometry to be a part of projective geometry, and projective geometry to be all of geometry.

Figure C.6 below gives a construction for perpendicular lines in H^2. Otherwise one must exercise caution with the Klein model, since angles and lengths are not faithfully represented. For example, a hyperbolic line has in fact infinite length although it is represented by a finite chord in the Klein model. See [Ryan] for a faithful model of H^2.

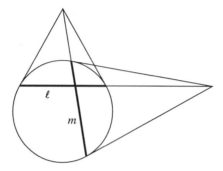

Figure C.6. Perpendicular lines ℓ and m in the Klein model.

Exercises

C.1   Prove that for vectors u, v, w in $\mathbf{R}^3$:

$$(u \times v) \times w - u \times (v \times w) = (u \cdot v)w - (v \cdot w)u.$$

Hint: Exercise 7.14 and associativity of quaternionic multiplication. Since triple product of vectors is a 3×3 determinant, $(u \times v) \cdot w = u \cdot (v \times w)$.

C.2   Use the previous exercise to prove that

$$(u \times v) \times w = (u \cdot w)v - (v \cdot w)u.$$

C.3   Use the preceding exercise to prove that

$$(u \times v) \cdot (v \times w) = (u \cdot v)(v \cdot w) - (v \cdot v)(u \cdot w).$$

C.4   Compute the area of a lune of angle α on the sphere, i.e. the shaded region in the figure. Show that the total area of the elliptic plane is 2π.
        Hint: Double integration with spherical coordinates.

Figure C.7. Lune.

C.5   Show that the area of an elliptic triangle with internal angles α, β, γ is $\alpha + \beta + \gamma - \pi$.

Hint: Use the preceding exercise and the partitioned lunes in the figure below.

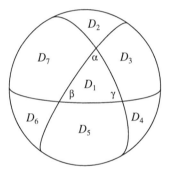

Figure C.8. Partitioned lunes.

C.6   Show that the altitudes of an elliptic triangle are concurrent. Do the same with the medians and angle bisectors of an elliptic triangle.

Hint: Use Theorem 11.13.

Appendix D

Ternary Rings

In Chapter 9 we did essentially the following: starting with an affine plane satisfying the minor and major Desargues' axioms, A4 and A5, respectively, and choosing a line with two preferred points, 0 and 1, we proved the nine axioms of a division ring to be satisfied by the points on ℓ with operations $\cdot$ and $+$. Coordinatizing the projective plane with Axiom P5 by a division ring turned out later to be a detail in this campaign.

In this appendix we are going to do the same thing to an affine plane *without* Axioms A4 and A5. We can no longer resort to translations and central dilatations for our definitions of addition and multiplication. We will see what algebraic conditions Axioms A1–A3 lead to a ternary ring — an algebraic structure on the set of points on a "diagonal" line, consisting of a ternary operation with several properties and from which limited notions of addition and multiplication may be recovered. Conversely, ordered pairs of coordinates from a ternary ring with lines defined by the ternary operation form an affine plane. Then we take up the interesting question of what extra conditions on the ternary ring of an affine plane Axiom A4 alone imposes. Indeed, as we add stronger axioms that converge to A5, we expect to see the ternary ring converge to a division ring.

The project we suggest in this appendix is to expand on what is presented and experiment with geometric axioms and their algebraic conditions on a ternary ring, or the converse. Also develop the corresponding projective geometry. You might add to your project by doing a small amount of scholarly investigation in the literature. The ternary ring is due to the 20th century American mathematician Marshall Hall: our source is [Blumenthal].

D.1   Coordinatization

Let **A** be an affine plane. (We are assuming only Axioms A1, A2, A3 listed in Chapter 1.) We proceed to give **A** coordinates from a ternary ring in several steps.

(*1*) Select any point in **A** and label it *O*: call *O* the *origin*.

(*2*) Using Axiom A3 and A2, we can show that there exist three distinct lines through *O*. Choose one and call it the *x-axis*, another to call the *y-axis*, and a third to call the *diagonal*.

(*3*) On the diagonal select a point *I* different from *O*. Refer to *I* as the *unit point*.

(*4*) Let Γ be an abstract set in one-to-one correspondence with the set of points on the diagonal. We adopt the convention that points are given capital letters, and, if the point is on *OI*, the corresponding element in Γ is the same letter in small case. The two exceptions are *O* and *I* to which we make $0, 1 \in \Gamma$ correspond.

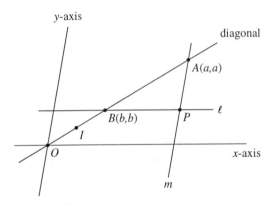

Figure D.1. Assigning coordinates to *P*. ($\ell \parallel$ *x*-axis, $m \parallel$ *y*-axis.)

(*5*) Let *A* denote any point on the diagonal, *OI*. We assign to *A* the coordinates (a,a): to *O* and *I*, we assign $(0,0)$ and $(1,1)$, respectively.

(*6*) Let *P* be a point in **A**, not on *OI*. By A2 there is a unique line through *P* and parallel to the *x*-axis. This line must intersect *OI* at some point, let us say *B*. Again, there is a line through *P* and parallel to the *y*-axis: this must intersect the diagonal at a point *A*. Assign to *P* the coordinate (a,b). This is an unequivocal assignment of coordinates in a bijective correspondence between $\Gamma \times \Gamma$ and **A**.

(*7*) We assign a slope and *y*-intercept to each line ℓ not parallel to the *y*-axis. Call the unique line through *I* and parallel to the *y*-axis the *line of slopes*. Let ℓ' be parallel to ℓ and passing through *O*. ℓ' will intersect the line of slopes, say at $(1,m)$. We assign to ℓ the *slope* $m \in \Gamma$. In addition, ℓ will intersect the *y*-axis, say at the point $(0,b)$: *b* is called the *y-intercept* of ℓ. Note that *b* and *m* uniquely determine a line through $(0,b)$ and parallel to the join of O and $(1,m)$, indeed, the line of slope m and y-intercept b.

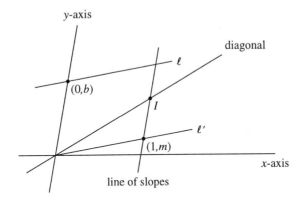

Figure D.2. Slope and y-intercept of a line in **A**.

(8) To each ordered triple, $(a,m,b) \in \Gamma \times \Gamma \times \Gamma$, we are going to assign a unique element $T(a,m,b)$ of Γ. This will define a ternary ring (Γ, T), i.e. a set Γ with ternary operation $T: \Gamma \times \Gamma \times \Gamma \to \Gamma$ satisfying five properties, T1–T5 below.

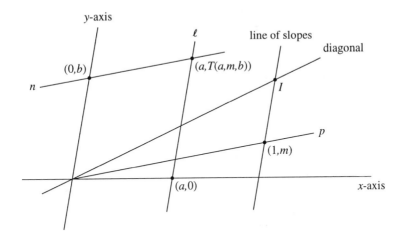

Figure D.3. Defining the ternary operation. $(p \parallel n.)$

The definition of $T(a,m,b)$ is very simple. Consider the line with slope m and y-intercept b: call it n. Let ℓ be the line through $(a,0)$ and parallel to the y-axis. Since n has a slope it is not parallel to the y-axis and will intersect ℓ at a point: the point will have coordinates of the form (a,y). Assign $T(a,m,b) = y$.

Note that an arbitrary point $P(x,y)$ lies on n if and only if the equation $y = T(x,m,b)$ is satisfied. Thus $y = T(x,m,b)$ is an equation of the line with slope m and y-intercept b. (A line parallel to the y-axis has equation of the form $x = a$.)

D.1. Definition A (planar) *ternary ring* (Γ, T) is a set $\Gamma = \{0,1,a,b,c,\ldots\}$ together with a mapping $T \colon \Gamma \times \Gamma \times \Gamma \to \Gamma$ such that T1–T5 are satisfied:

> **T1.** For all $a,b,c \in \Gamma$,
> $$T(0,b,c) = T(a,0,c) = c.$$
>
> **T2.** For all $a \in \Gamma$,
> $$T(a,1,0) = T(1,a,0) = a.$$
>
> **T3.** If $m,m',b,b' \in \Gamma$ and $m \neq m'$, then the equation
> $$T(x,m,b) = T(x,m',b')$$
> has a unique solution in Γ.
>
> **T4.** If $a,a',b,b' \in \Gamma$, $a \neq a'$, the system of equations
> $$T(a,x,y) = b$$
> $$T(a',x,y) = b'$$
> has a unique solution.
>
> **T5.** For all $a,m,c \in \Gamma$, the equation
> $$T(a,m,x) = c$$
> has a unique solution.

D.2. Example The set Γ with ternary operation defined in steps *(1)*–*(8)* is a ternary ring. T1 and T2 are special cases for the lines n or ℓ. T3 is equivalent to the proposition that lines are either parallel (and have the same slope) or intersect in precisely one point. T4 is equivalent to Axiom A1: through the distinct points (a,b) and (a',b') there is one and only one line ℓ of slope x and intercept y. T5 is equivalent to Axiom A2: there is a unique line through (a,c) and parallel to OM where $M = M(1,m)$. Our use of the word "equivalent" is not an accident: a ternary ring (Γ, T) defines on $\Gamma \times \Gamma$ an affine plane with lines given by $\{(x,y) \mid x = a\}$ and $\{(x,y) \mid y = T(x,m,b)\}$ (for all $m,b \in \Gamma$). You will then be able to prove A1–A3 by reversing the reasoning above.

D.3. Example A division ring $(R,+,\cdot,0,1)$ is a ternary ring. Define $T(a,m,b) = a \cdot m + b$. You should now check that properties T1–T5 are satisfied.

We continue by defining some limited notions of addition and multiplication.

D.2 Addition

Given a ternary operation T, we define a binary operation $\Gamma \times \Gamma \to \Gamma$ we call *addition* and denote it simply by $(a,b) \mapsto a+b$. Define

$$a+b = T(a,1,b).$$

Figure D.4 below indicates how addition works on the diagonal.

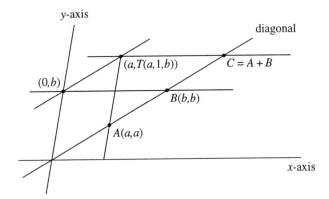

Figure D.4. Addition of diagonal points.

D.4. Definition A *loop* is a set Λ with preferred element 0, and binary operation $\odot$ such that properties L1 and L2 are satisfied.

> **L1.** $a \odot 0 = a = 0 \odot a$ for all $a \in \Lambda$
> **L2.** $a \odot b = c$ uniquely determines any one of $a,b,c \in \Lambda$ whenever the other two are given.

D.5. Remark There is a slight redundancy in the definition of loop: where is it? Examples of a loop are $(\Gamma,+,0)$ and any group $(G,\cdot,e)$.

Although $(\Gamma,+,0)$ is a loop it is generally neither true that it is a group nor that $+$ is a commutative operation. It is a fact that in the presence of the minor Desargues' axiom $(\Gamma,+,0)$ is an abelian group.

D.3 Multiplication

Given a ternary operation T, we define a binary operation on Γ called *multiplication* and denoted by $(a,b) \mapsto a \cdot b$. Define

$$a \cdot b = T(a,b,0).$$

Figure D.5 below gives the resulting construction of multiplication on the diagonal.

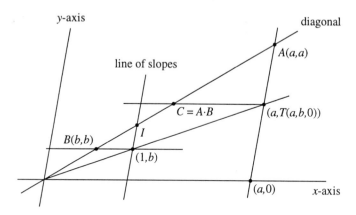

Figure D.5. Multiplication of diagonal points.

Now show that

(*1*) $(\Gamma \smallsetminus \{0\}, \cdot, 1)$ is a loop.
(2) $A \cdot B = 0$ if and only if $A = 0$ or $B = 0$ $(A, B \in OI)$.

D.6. Remark We now have simple equations for certain lines in **A**: $y = x + b$, $y = x \cdot m$, $y = b$, $x = a$. (Notice the change in left-right convention from Chapters 8 and 9.) It is in general not true that $T(x, m, b) = x \cdot m + b$. This would be equivalent to the equation

$$T(x \cdot m, 1, b) = T(x, m, b).$$

If such is the case for all ordered triples, T is said to be *linear*. Another outcome of assuming A4 is that T is linear.

The next theorem requires a brief definition and example to precede it. You will be required to supply its proof with the help of the Exercises D.2–D.4.

D.7. Definition A ternary ring (Γ, T) is called a *Veblen-Wedderburn system* if and only if

VW1. $(\Gamma, +, 0)$ is an abelian group.
VW2. $(\Gamma \smallsetminus \{0\}, \cdot, 1)$ is a loop.
VW3. $a \cdot 0 = 0 \cdot a = 0$ for all $a \in \Gamma$.
VW4. Right distributivity: $(a + b) \cdot c = a \cdot c + b \cdot c$ for all $a, b, c \in \Gamma$.

D.8. Example Consider an 8-dimensional real vector space with basis $e_1, \ldots, e_8$, with $e_1 = 1$. We will define the nonassociative algebra of *octonions* **O** on this vector space. Introduce multiplication on the basis elements by letting $1e_i = e_i = e_i 1$ for $i = 1, \ldots, 8$, $e_i^2 = -1$ for $i \geq 2$, and $e_i e_j = -e_j e_i$ for $1 < i < j \leq 8$. We also let

$$e_2 e_3 = e_4, \qquad e_2 e_5 = e_6, \qquad e_3 e_5 = e_7, \qquad e_4 e_5 = e_8,$$
$$e_6 e_4 = e_7, \qquad e_7 e_2 = e_8, \qquad e_8 e_3 = e_6,$$

with 14 more relations gotten by cyclically permuting indices, i.e. if $e_i e_j = e_k$ we require $e_j e_k = e_i$ and $e_k e_i = e_j$. Multiplication between general octonions is obtained by using both distributive laws and now letting the real numbers commute. The following is a mnemonic scheme for octonionic multiplication.

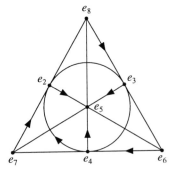

Figure D.6. Diagram for octonionic multiplication.

The octonions are an example of a Veblen-Wedderburn system. (Why?)

D.9. Theorem *An affine plane* **A** *satisfies A4, the minor Desargues' axiom, if and only if* **A** $\cong \Gamma \times \Gamma$, *where* (Γ, T) *is a Veblen-Wedderburn system, and lines are given by* $x = a$ *and* $y = T(x, m, b)$.

Exercises

D.1   Give a synthetic proof for the existence of an affine plane of nine points $A, B, C, \ldots, I$. Coordinatize with $\Gamma = \{0, 1, 2\}$. Assign slopes to each line ABC, etc. Evaluate $T(2, 2, 2)$, $T(1, 2, 1)$ and $T(2, 1, 1)$. Write equations for several lines. In this example, does $T(x, m, b) = x \cdot m + b$?

D.2   Does the Moulton plane satisfy A4?

Define a *vector* in **A** by an ordered pair $\overline{AB}$. Line AB is the carrier of the vector

$\overline{AB}$. If $A = B$, $\overline{AB}$ is a *null*. Define a relation $\simeq$ among vectors as follows:

$$\overline{AB} \simeq \overline{CD} \quad \Longleftrightarrow \quad \begin{cases} A = B \text{ and } C = D, \text{ or} \\ A = C \text{ and } B = D, \text{ or} \\ AB \parallel CD \text{ and } AC \parallel BD. \end{cases}$$

D.3   Show that $\simeq$ is a symmetric and reflexive relation. Show $\simeq$ is a transitive relation if and only if A4 holds in **A**.

Let $\overline{AB}$ and $\overline{CD}$ be two vectors (now assumed to be equivalence classes of $\simeq$). Define an addition as follows: if $B = C$, $\overline{AB} + \overline{CD} = \overline{AD}$. If $B \neq C$, and Q denotes the unique point such that $\overline{BQ} \simeq \overline{CD}$ (in the presence of A4), then $\overline{AB} + \overline{CD} = \overline{AQ}$.

D.4   Show that the set of vectors is an abelian group under $+$. Show by these means, or by means of translations, that $(\Gamma, +, 0)$ is an abelian group.

D.5   Show that, in the presence of A4, the ternary operator T is linear.

D.6   Show that, in the presence of A4, multiplication is right distributive over addition. Complete the remaining steps in the proof of Theorem D.9.

D.7   Let **F** be the field of nine elements in Exercise 7.12. Let **A** = **F** as sets. Prove that **A** is a Veblen-Wedderburn system if the ternary operation is defined by

$$T(a,m,b) = \begin{cases} am + b & \text{if } m \text{ is a square in } \mathbf{F} \\ a^3 m + b & \text{if } m \text{ is not a square in } \mathbf{F}. \end{cases}$$

The right-hand side should be interpreted as operations in the field **F**. A square y in **F** has $z \in \mathbf{F}$ such that $y = z^2$. Prove that $(\mathbf{A} \setminus \{0\}, \cdot, 1)$ is in fact a group.

D.8   Define a projective plane over **A** in the preceding exercise; denote it by $\mathbf{P}_\mathbf{A}^2$. Show that Desargues' theorem does not hold in $\mathbf{P}_\mathbf{A}^2$: an example of a finite non-Desarguesian plane.

D.9   Give coordinates to a general projective plane.

Appendix E

The Lattice of Subspaces

Our experience with the incidence geometry of projective planes has taught us that join and intersection are the two fundamental operations; alternatively, the relation of set inclusions among the points and lines also determines the geometry. For example, that three points A, B, C are collinear may either be stated by $A \cup B = B \cup C$ or by $A \in BC$. The operations of join and intersection are easily extended to binary operations on the set of points, lines, the whole plane and the empty set: the set of these components, called subspaces, of the projective plane, together with join and intersection, and the basic rules of interaction they satisfy, form an abstract algebraic system called a lattice. Lattices have eventually been observed in almost all of modern mathematics. Lattices also have a principle of duality associated with them, and it is by this extension of the principle of duality from geometry to algebra that duality has come to pervade a good part of mathematics. You need only think of the dual vector space to see that it has even had an effect on freshman mathematics.

In this appendix, we will introduce the reader to lattices, the general theory of synthetic projective geometry, the associated principle of duality and fundamental theorem of projective geometry. The synthetic theory will be so general that it will include any finite dimension or even infinite dimension as a special case. We will see that the extra territory is obtained at the expense of very little extra complication over the synthetic theory in Chapters 1–11. In this appendix, we also treat partially ordered sets with least upper bound and greatest lower bound as an alternative definition of lattice, which emphasizes the ordering one can make on the subspaces of projective geometry. Further reading is to be found in [Jacobson], [Holland], [von Neumann] and [Birkhoff].

E.1   Posets and Lattices

Recall that a (binary) relation on a set S is a subset $\mathfrak{R}$ of the Cartesian product $S \times S$. If $(a,b) \in \mathfrak{R}$, one often writes $a\mathfrak{R}b$. In the next definition, we use the symbol $\geq$ instead of $\mathfrak{R}$.

E.1. Definition A *partially ordered set* (poset) is a set S together with a relation $\geq$ such that the following conditions are met:

> **PO1.** Reflexivity: $a \geq a$ for every $a \in S$;
> **PO2.** Anti-symmetry: If $a \geq b$ and $b \geq a$, then $a = b$;
> **PO3.** Transitivity: If $a \geq b$ and $b \geq c$, then $a \geq c$.

A standard example of a partially ordered set is the power set of a set S (i.e., the set of all subsets including the empty set and S), or any other set of subsets of S under inclusion. Particular cases of this are the set of subgroups of a group, or the set of subspaces of a vector space, under set inclusion. Here is a seemingly unrelated example of a poset: on the natural numbers $\mathbf{N}$ place the partial ordering $a \geq b$ iff a divides b (denoted by $a|b$).

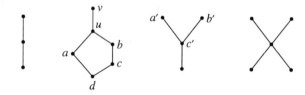

Figure E.1. Examples of finite posets, the one on the left being *well-ordered*.

Given two elements a,b in a poset S, their *least upper bound* (if it exists) is an element $u \in S$, denoted by $u = a \vee b$, such that two conditions are met:

> (1) $u \geq a$ and $u \geq b$;
> (2) If $v \geq a$ and $v \geq b$ for any other element $v \in S$, then $v \geq u$.

See the figure for an example utilizing the same letters: however, a' and b' in another poset diagrammed lack a least upper bound (indeed, any upper bound). Least upper bound (lub) corresponds to set union in the power set example: in this case, least upper bound of two elements always exists. In the natural number example, least upper bound of a and b is their greatest common divisor. Lastly, observe that if $a \vee b$ exists in a poset $(S, \geq)$, it is a uniquely determined element by condition PO2.

Dually, we define the *greatest lower bound* of $a,b \in S$ to be an element $w \in S$, $w = a \wedge b$ (if it exists) such that two conditions are met:

(1) $a \geq w$ and $b \geq w$;
(2) If $a \geq v$ and $b \geq v$ for any other element $v \in S$, then $w \geq v$.

In Figure E.1, we have $a' \wedge b' = c'$. Greatest lower bound (glb) corresponds to intersection in the power set example, and least common multiple in the natural number example.

E.2. Definition A *lattice* is a poset in which any two elements have a least upper bound and a greatest lower bound.

As brought out above, the power set of a set S is a lattice: if $X, Y \subseteq S$, then $X \vee Y = X \cup Y$ (ordinary union of subsets) and $X \wedge Y = X \cap Y$ (ordinary intersection). A key example of a lattice is the poset of all subspaces of a vector space V. If U and W are subspaces of V, the lattice operations $U \wedge W = U \cap W$ (intersection of subspaces: itself a subspace!) and $U \vee W = U + W$ (the sum of U and W, i.e., the subspace of linear combinations of vectors from U and W). Finally, exactly two of the finite posets in Figure E.1 are lattices: which are they?

You will be asked to check in Exercise E.4 that $a \wedge b$ and $a \vee b$ satisfy the four (times two) conditions L1–L4 below. In fact, these four conditions on a set L with binary operations $\vee$ and $\wedge$ characterize lattices. We are led to an *alternate definition of lattice.*

E.3. Definition A *lattice* is a set L together with two binary operations $\wedge$ and $\vee$ $L \times L \to L$ such that conditions L1–L4 are satisfied (for all $a, b, c \in L$):

L1. Commutativity: $a \vee b = b \vee a$ and $a \wedge b = b \wedge a$;
L2. Associativity: $(a \vee b) \vee c = a \vee (b \vee c)$ and $(a \wedge b) \wedge c = a \wedge (b \wedge c)$;
L3. Idempotency: $a \vee a = a$ and $a \wedge a = a$;
L4. Order: $(a \vee b) \wedge a = a$ and $(a \wedge b) \vee a = a$.

Again, you will be asked to check that a poset $(S, \geq)$, in which every pair of elements has a glb or lub, satisfies the conditions L1–L4 (Exercise E.3). Conversely, a set L with binary operation $\vee$ and $\wedge$ satisfying L1–L4 is a poset in which each pair of elements has a lub and glb. Define a relation $\geq$ on L by requiring that

$$a \geq b \qquad \Longleftrightarrow \qquad a \vee b = a,$$

or equivalently, $a \wedge b = b$. Note that condition PO1, i.e., $a \geq a$ for all $a \in L$, follows from L3. You are asked to check in Exercise E.3 that conditions PO2 and PO3 follow from L1–L4 as well. Hence, L together with this ordering $\geq$ is a poset. Finally, each pair of elements in L has a lub and glb, since, given $a, b \in L$, $a \vee b$

is their glb and $a \wedge b$ is their lub. In order to see this, we must verify the two required conditions. First, $a \vee b \geq a$ and $a \vee b \geq b$, since $(a \vee b) \vee a = a \vee b$ and $(a \vee b) \vee b = a \vee b$ by L1–L3. Second, given $v \in L$ such that $v \geq a$ and $v \geq b$, i.e. $v \vee a = v$ and $v \vee b = v$, then $(a \vee b) \vee v = v \vee (a \vee b) = (v \vee a) \vee b = v \vee b = v$, hence $v \geq a \vee b$. Hence, $a \vee b$ is the least upper bound of a and b. Similarly, $a \wedge b$ is the glb of a and b (Exercise E.3).

Since the conditions L1–L4 in the second definition of lattice state precisely the same assumptions for the binary operation $\vee$ as for the binary operation $\wedge$, we clearly have the following principle.

E.4. The Principle of Duality for Lattices *If S is a statement about lattices which is deduced from L1–L4, then the dual statement S′ obtained by interchanging $\vee$ and $\wedge$ is deducible from L1–L4 as well.*

Indeed, the complete proof for S, with $\vee$ and $\wedge$ interchanged, supplies a proof for S'.

For example, we have just shown that given lattice $(L, \vee, \wedge)$, satisfying L1–L4, the $\geq$ relation defined by $a \geq b$ iff $a \vee b = a$ (iff $a \wedge b = b$) makes L a poset. The dual statement is that $(L, \geq')$ is a poset, where

$$a \geq' b \quad \Longleftrightarrow \quad a \wedge b = a.$$

This is automatically true by the principle of duality, but, if pressed to supply a proof, one might interchange $\wedge$ and $\vee$ in the argument supplied above and in Exercise E.3. (Incidentally, $a \geq b$ iff $b \geq' a$ as one may check. One says that the dual poset of poset $(S, \geq)$ is $(S, \geq')$, where $a \geq' b$ iff $b \geq a$. Check that $(S, \geq')$ is itself a poset (check PO1–PO3). What is the dual lattice $(L, \vee', \wedge')$ of $(L, \vee, \wedge)$?)

E.5. Definition A *homomorphism of lattices* $(L, \vee, \wedge)$ and $(L', \vee, \wedge)$ is a mapping $T: L \to L'$ which satisfies $T(a \wedge b) = T(a) \wedge' T(b)$ and $T(a \vee b) = T(a) \vee T(b)$ (for all $a, b \in L$). A lattice homomorphism is *order-preserving*, which means that if $a \leq b$, then $T(a) \leq' T(b)$ (Exercise E.5). A bijective lattice homomorphism is called an *isomorphism* and the two lattices so related are said to be *isomorphic*.

A lattice L may contain a greatest element 1, where $1 \wedge a = a$ (i.e., $1 \geq a$) for all $a \in L$. L may also contain a least element 0, where $0 \vee a = a$ (i.e., $a \geq 0$) for all $a \in L$. (Prove that 0 and 1 are unique.) A finite well-ordered set provides an example of a lattice with 0 and 1.

E.6. Definition A *complemented lattice* is a lattice L with 0 and 1 such that for each $a \in L$ there exists an $a' \in L$ such that $a \vee a' = 1$ and $a \wedge a' = 0$. a' is called the *complement* of a, and is usually not unique.

For example, the lattice of all subspaces $L(V)$ of a finite dimensional vector space V is a complemented lattice, since given any subspace U of V, a basis for U may be extended to a basis for V, the extra basis vectors spanning a complement (complementary subspace). The process of completing to a basis of V is by no means unique.

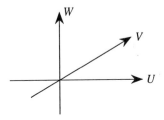

Figure E.2. Two complements of the one-dimensional subspace U in a two-dimensional vector space.

E.2   Axioms for General Projective Geometry

We will now give three simple axioms for projective geometry (O. Veblen and J.W. Young, 1919). We then sketch a synthetic theory of subspace and dimension, extract a lattice of subspaces and state a most general form of the fundamental theorem of projective geometry. The student project intended here should center around understanding the fundamental theorem, its relation to the theorems of the same name in Chapters 1–11, and a study of its proof in the various sources given.

E.7. Definition A *projective geometry* **P** is a set with at least 2 elements, whose elements are called *points*, together with a set of subsets of **P**, whose elements are called *lines*, such that the three axioms below are met.

> **PG1.** For every two points in **P** there is one, and only one, line that contains both these points.
> (Notation: if P and Q are points in **P**, denote by $P \vee Q$ the unique line containing ("passing through") P and Q. We also set $P \vee P = P$ for each point P.)
>
> **PG2.** Every line has at least three points.
> (Definition: A *triangle* consists of three non-collinear points P, Q, R, and the three lines $P \vee Q$, $Q \vee R$, and $P \vee R$, called *sides*.)
>
> **PG3.** If a line ℓ intersects two sides of a triangle at different points, then ℓ intersects the third side of a triangle.

Axiom PG3 is clearly the non-Euclidean axiom corresponding to P2 in Chapter 2.

E.8. Definition   A subset $\ell \subseteq \mathbf{P}$ is called a *subspace* if

$$P, Q \in \ell \qquad \Longrightarrow \qquad P \vee Q \subseteq \ell.$$

The empty set is a subspace denoted by 0, while all of the point set $\mathbf{P}$ is a subspace denoted by 1. It is clear from $P \vee P = P$ that a point by itself is a subspace. A subspace ℓ together with the lines contained in it is a projective geometry in its own right denoted by $[0, \ell]$.

Given two subspaces ℓ_1 and ℓ_2, define their binary composition by

$$\ell_1 \vee \ell_2 = \{X \in \mathbf{P} \mid X \in P \vee Q \text{ for some } P \in \ell_1, Q \in \ell_2.\}$$

If $\ell_1 = 0$, we let $0 \vee \ell_2 = \ell_2 = \ell_2 \vee 0$. It is not hard to show that $\ell_1 \vee \ell_2$ as defined is indeed a subspace, and that $\vee$ is associative. Also, $\vee$ is clearly commutative and idempotent.

A binary operation on the set of subspaces of $\mathbf{P}$ is defined by

$$\ell_1 \wedge \ell_2 = \ell_1 \cap \ell_2.$$

The intersection of two subspaces is easily verified to be a subspace: just as easy it is to show (in Exercise E.10) that the set of subspaces $L(\mathbf{P})$ of a projective geometry $\mathbf{P}$, together with $\vee$ and $\wedge$ as defined above, is a modular lattice, where the partial ordering $\geq$ is just set inclusion $\supseteq$.

Given a finite set of points $\{P_1, \dots, P_n\}$ in a projective geometry $\mathbf{P}$, we say that these are *independent* if

$$(P_{i_1} \vee \cdots \vee P_{i_k}) \wedge (P_{i_{k+1}} \vee \cdots \vee P_{i_n}) = 0$$

for all permutations $j \mapsto i_j$, $j = 1, \dots, n$ of n letters. A similar definition of independence may be given for an infinite family of points (cf. [Holland]). If the points $\{P_1, \dots, P_n\}$ furthermore generate $\mathbf{P}$, i.e., $\mathbf{P} = P_1 \vee \cdots \vee P_n$, we say they form a *basis* for $\mathbf{P}$. Every independent set of points forms a basis for the subspace $\ell = P_1 \vee \cdots \vee P_n$ they generate. The number $n - 1$ is defined to be the dimension of the subspace ℓ. For example, four points never will be independent in a projective plane, but three noncollinear points are so. If $\mathbf{P}$ has an infinite basis, $\mathbf{P}$ is said to have infinite dimension, denoted by $\dim(\mathbf{P}) = \infty$. For any subspaces ℓ, m of a projective geometry $\mathbf{P}$, we have the formula familiar from vector space theory:

$$\dim(\ell \vee m) + \dim(\ell \wedge m) = \dim(\ell) + \dim(m),$$

where the intuitive operations with infinity are valid (such as $\infty + a = \infty$).

E.9. Definition An *isomorphism*, or *projectivity*, $\phi\colon \mathbf{P}_1 \to \mathbf{P}_2$ of projective geometries is a one-to-one correspondence of the underlying point sets such that

$$\phi(P \vee Q) = \phi(P) \vee \phi(Q) \qquad \text{(for all } P, Q \in \mathbf{P}_1\text{)}.$$

In Exercise E.11, you may verify that ϕ induces a lattice isomorphism $L(\mathbf{P}_1) \to L(\mathbf{P}_2)$ (and order-preserving mapping, so that if $\ell \subseteq m$, then $\phi(\ell) \subseteq \phi(m)$).

Given subspaces ℓ and m of $\mathbf{P}$, where $\dim(\ell) \geq 1$, and a third subspace p such that $\ell \vee p = m \vee p$ and $\ell \wedge p = m \wedge p = 0$, then the mapping $\phi\colon [0, \ell] \to [0, m]$ given by

$$\phi(Q) = (Q \vee p) \wedge m \qquad (Q \in \ell)$$

is an isomorphism of projective geometries called a *perspectivity with axis p*. A perspectivity $\ell \overset{p}{\underset{\wedge}{=}} \ell'$ in the projective plane is clearly an example.

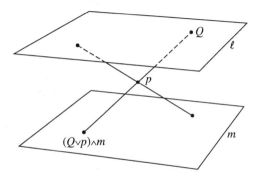

Figure E.3. A perspectivity with axis p.

Given a subspace ℓ of the projective geometry $\mathbf{P}$, a subspace m which satisfies $\ell \vee m = 1$ and $\ell \wedge m = 0$ is a complement of ℓ. Every subspace has a complement, although it is by no means unique. For example, a complement to the line ℓ in a projective plane is any point $X \notin \ell$. The dimension of a complement to a subspace ℓ is called the *codimension* of ℓ. A *hyperplane* is a subspace of codimension 0: perhaps it is more aptly called a "dual-point", since the principle of duality extends to the interchange of points and hyperplanes in this theory (Exercise E.13).

E.10. Definition (G. Birkhoff) A *polarity* on a projective geometry $\mathbf{P}$ is a mapping $\ell \mapsto \ell^{\perp}$ of $L(\mathbf{P})$ into itself such that

   PL1. $1^{\perp} = 0$;
   PL2. $\ell \leq m$ implies $\ell^{\perp} \geq m^{\perp}$;
   PL3. If P is a point, then $P^{\perp}$ is a hyperplane and $P^{\perp\perp} = P$.

$P^{\perp}$ is called the *polar* of P, and P the *pole* of $P^{\perp}$.

It follows from the three conditions that $\ell \subseteq \ell^{\perp\perp}$ for every subspace ℓ. If **P** is a two-dimensional geometry, we recover the polarity defined in Appendix A and Chapter 3. In fact, so long as $\dim \mathbf{P} < \infty$, a polarity is an isomorphism of the lattice $L(\mathbf{P})$ with its dual lattice.

If m is a subspace of a projective geometry **P**, the *relative polar* induced on m is defined on a subspace $\ell \leq m$ by $\ell \mapsto \ell' = \ell^{\perp} \cap m$. Check that this is a polar on $[0,m]$.

A polarity $\perp$ is said to be *orthomodular* if $\ell \vee \ell^{\perp} = 1$ for each subspace ℓ with $\ell = \ell^{\perp\perp}$ (see [Holland] for several equivalent conditions for orthomodular polarity). In particular, an orthomodular polar $\perp$ is not a hyperbolic polarity, and there will be no nonempty conics associated with $\perp$ (cf. Appendix A).

E.3   The Fundamental Theorem of Projective Geometry

The primary example of a projective geometry is built from a division ring R and a right vector space V over R (i.e., all axioms of a vector space are satisfied except $\alpha v = v\alpha$). Suppose the dimension of V is 2 or more, perhaps infinite dimensional. The points of $\mathbf{P}(V,R)$, the associated projective geometry, are the 1-dimensional subspaces of V, while a line is the family of 1-dimensional subspaces contained in a 2-dimensional subspace. Then the three axioms PG1–PG3 are satisfied by $\mathbf{P}(V,R)$ (Exercise E.15). The subspaces of $\mathbf{P}(V,R)$ correspond precisely to subspaces of V, although dimension of a subspace is one more than its projective dimension. The next theorem asserts in its first part the universality of this example in dimension 3 or more.

E.11. The Fundamental Theorem *Existence Part: Given a projective geometry* **P** *of dimension $n \geq 3$ (or ∞), there is a division ring R and a right vector space V over R such that* **P** *is isomorphic to* $\mathbf{P}(V,R)$.

Uniqueness Part: Suppose V is a right vector space over a division ring R, where $3 \leq \dim(V) \leq \infty$, and W is a right vector space over a division ring S such that there exists a projectivity $\phi \colon \mathbf{P}(V,R) \to \mathbf{P}(W,S)$. Then there exists a ring isomorphism $\sigma \colon R \to S$ and an invertible semi-linear transformation $T \colon V \to W$ (i.e., $T(x\lambda) = T(x)\sigma(\lambda)$ and $T(x+y) = T(x) + T(y)$ for all $\lambda \in R$ and $x,y \in V$) such that $\phi(P) = Tx$ where $P = xR$. (One says that T implements ϕ.)

The existence part should be quite plausible to readers of this book, since $\dim \mathbf{P} \geq 3$ gives us Desargues theorem (Theorem 3.1); then we have the coordinatization of Chapter 9 for Desarguesian planes. [von Neumann] has a proof of the existence part for the more general case of complemented, modular lattices, and the reader may obtain a proof from Part II, Theorem 14, or from a sketch of von Neumann's method in [Holland].

We have proven the uniquness part of the fundamental theorem in the case $\dim V = 3$ and $V = W$, $R = S$. The technique of proof followed in Chapter 8 is easily extended to any finite dimension. The full uniqueness theorem is proven in [Jacobson], who in fact proves that a lattice isomorphism

$$(L(V_R), \vee, \wedge) \to (L(W_S), \vee, \wedge)$$

is implemented by a semi-linear transformation, which is a more general result.

As a footnote to our discussion of 20th century versions of the 19th century fundamental theorem (of von Staudt), we mention how polarity coordinatizes. A polarity $\perp$ on a projective geometry always arises as a bilinear form $\langle\,,\,\rangle : V \times V \to R$ on the vector space V over R, where $\mathbf{P} \cong \mathbf{P}(V,R)$. However, this bilinear form might well be degenerate, i.e., we could find $x \neq 0$ such that $\langle x, x \rangle = 0$. This is part of the statement of the Birkhoff–von Neumann theorem (cf. [Holland]). Orthomodularity for $\perp$ and for $\langle\,,\,\rangle$ have precisely the same definition: for every subspace U, $U + U^{\perp} = V$ where $U^{\perp} = \{x \in V \mid \langle x, u \rangle = 0 \text{ for all } u \in U\}$. In particular, orthomodular bilinear forms are not degenerate.

The next theorem is a result from 1994 in [Holland]. We will need the notion of harmonic conjugate in a projective geometry $\mathbf{P}$ of dimension 2 or more.

E.12. Definition Let P and Q be two distinct points of $\mathbf{P}$, and C a third point on the line $P \vee Q$. The *harmonic conjugate* of C with respect to P and Q is a fourth point D obtained as follows: choose a point $X \notin P \vee Q$ and a third point Y on $X \vee P$, then construct the points $W = (C \vee Y) \wedge (Q \vee X)$ and $Z = (P \vee W) \wedge (Q \vee Y)$. The desired point is $D = (X \vee Z) \wedge (P \vee Q)$.

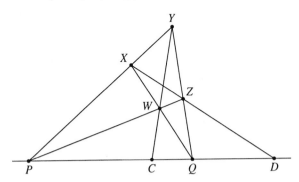

Figure E.4. Construction of the harmonic conjugate of C with respect to P and Q.

For example, if $\dim \mathbf{P} = 2$, then P and Q are the diagonal points of the complete quadrangle $WXYZ$ constructed, while $C = WY.PQ$ and $D = XZ.PQ$.

Before stating the theorem below, which can be said to characterize real, complex and quaternionic projective geometry in terms of some simple axioms, we will need to know that points $\{\ldots, P_i, \ldots, P_j, \ldots\}$ are *orthogonal* in a projective geometry $\mathbf{P}$ with polarity $\perp$ if $P_i \in P_j^{\perp}$ for every $i \neq j$. The details of inner product space may be found in [Holland].

E.13. Theorem (Holland) *Suppose* $\mathbf{P}$ *is a projective geometry with orthomodular polarity* $\perp$ *such that there is a sequence of orthogonal points* $\{P_i \mid i = 1, 2, \ldots\}$ *where every line* $P_i \vee P_{i+1}$ *contains a third point* C_i *with relative polar coinciding to the harmonic conjugate with respect to* P_i *and* P_{i+1}. *Then there exists a real, complex or quaternionic conjugate-bilinear inner product space* $(V, \langle \, , \, \rangle)$ *such that* $\mathbf{P} \cong \mathbf{P}(V, \mathbf{F})$ *(where* $\mathbf{F} = \mathbf{R}, \mathbf{C}$ *or* $\mathbf{H}$*) and the polarity coincides with the association of orthogonal subspace to a subspace.*

Exercises

E.1   Using at most a five element set S, realize each of the diagrams in Figure E.1 as the graph of a poset of subsets of S.

E.2   Draw poset diagrams, as in Figure E.1, for the following posets.
   (*a*) The power set of $\{1, 2, 3\}$.
   (*b*) $\{1, 2, 3, 6\}$ under divisibility.
   (*c*) The two-dimensional vector space $\mathbf{Z}_2 \times \mathbf{Z}_2$ (over the field $\mathbf{Z}_2$) and its subspaces.

E.3   Show that Definitions E.2 and E.3 of lattices are equivalent.
   Hint: Refer to the discussion following Definition E.3.

E.4   (*a*) Draw diagrams of the dual posets to the finite posets in Figure E.1. Are any *self-dual*, i.e. isomorphic to their duals?
   (*b*) The dual vector space of a vector space V over a field $\mathbf{F}$ is the set of all linear transformations $\phi \colon V \to \mathbf{F}$ with pointwise addition and scalar multiplication. I.e., $(\phi + \psi)(v) = \phi(v) + \psi(v)$ and $(\alpha\phi)(v) = \alpha(\phi(v))$. Denote this vector space by V^*. If $\{e_1, \ldots, e_n\}$ is a basis of V, then $\{e_1^*, \ldots, e_n^*\}$ is a basis for V^*, where e_i^* is defined by $e_i^*(\alpha_1 e_1 + \cdots + \alpha_n e_n) = \alpha_i$.
   Show that the lattice of all subspaces of V — denote this lattice by $L(V)$, where $U \vee W = U + W$ and $U \wedge W = U \cap W$ — is isomorphic to the lattice $(L(V^*), \wedge', \vee')$, where $U \wedge' W = U + W$ and $U \vee' W = U \cap W$. If V is finite dimensional, show that $(L(V), \geq)$ is self-dual.
   Hint: Try the mapping $U \mapsto U^* = \{\phi \in V^* \mid \phi(X) = 0 \text{ for all } x \in U\}$. Show that if $U \subseteq W$, then $U^* \supseteq W^*$.

E.5   A mapping $T \colon S \to S'$ between posets $(S, \geq)$ and $(S, \geq')$ is called *order-preserving* if $a \geq b$ implies $T(a) \geq' T(b)$.

(a) Show that a lattice homomorphism $\phi: (L, \vee, \wedge) \to (L, \vee', \wedge')$ induces an order-preserving mapping of posets $\hat{\phi}: (L, \geq) \to (L', \geq')$, where $a \geq b$ iff $a \vee b = a$, and $c \geq' d$ iff $c \vee' d = c$.

(b) Show: T is a lattice isomorphism iff T and its inverse are order-preserving.

E.6 (a) Show that one of the distributive laws in a lattice implies the other.

(b) Show that a well-ordered poset is a distributive lattice.

E.7 Let V be a vector space over a field $\mathbf{F}$. Let $L(V)$ be the lattice of all subspaces of V. Prove that $L(V)$ is a modular lattice.

E.8 If a and b are elements of a modular lattice, then the mapping given by $x \mapsto x \wedge b$ is a lattice isomorphism of the intervals $I[a, a \vee b]$ and $I[a \wedge b, b]$. The inverse is given by $y \mapsto y \vee a$. Prove.

E.9 Intervals $I[x, y]$ and $I[z, w]$ of a modular lattice are called *transposes* if there exist $a, b \in L$ such that $I[x, y] = I[a, a \vee b]$ and $I[z, w] = I[a \wedge b, b]$. By Exercise E.8, the two intervals are isomorphic. Two intervals $I[u, v]$ and $I[r, s]$ are called *projectively equivalent* if there exists a finite sequence of intervals of which each interval is a transpose of its predecessor. Show that projective equivalence is an equivalence relation among intervals.

E.10 Given a projective geometry $\mathbf{P}$, check that the set $(L(\mathbf{P}), \vee, \wedge, 0, 1)$ of all subspaces of $\mathbf{P}$, is a complemented modular lattice.

E.11 Given a projectivity $\phi: \mathbf{P}_1 \to \mathbf{P}_2$ and subspaces ℓ and m of $\mathbf{P}$, show that
(a) $\phi(\ell)$ is a subspace;
(b) $\ell \subseteq m$ iff $\phi(\ell) \subseteq \phi(m)$;
(c) $\phi(\ell \vee m) = \phi(\ell) \vee \phi(m)$;
(d) $\phi(\ell \wedge m) = \phi(\ell) \wedge \phi(m)$.

E.12 In a three-dimensional projective geometry, let ℓ and m be different two-dimensional subspaces. Given an ordered quadruple of points in general position on each of ℓ and m, find a sequence of three perspectivities carrying one ordered quadruple into the other.

E.13 Given a projective geometry $\mathbf{P}$ of dimension n, let $\mathbf{P}^*$ be the family of hyperplanes in $\mathbf{P}$ together with a set of distinguished subsets consisting of pencils of hyperplanes at each subspace of dimension $n - 2$. Show that $\mathbf{P}^*$ is itself a projective geometry of dimension n. Formulate a principle of duality as in Chapter 3.

E.14 Prove the theorem of Desargues: Let p and p' be two different subspaces of finite dimension ≥ 2 in a projective geometry. Let ℓ be a subspace of p, $\ell' \subseteq p'$ be its image space under a perspectivity $[0, p] \to [0, p']$. Then $\ell \wedge \ell' = \ell \wedge (p \wedge p')$. Dualize.

E.15 Given a division ring R and right vector space V over R, show that $\mathbf{P}(V, R)$ satisfies conditions PG1–PG3.

E.16 Prove the converse of Theorem E.13.

Solutions to Selected Exercises

2.2 (*a*) Define a bijective assignment of lines in **S** (other than ℓ) into lines of $\mathbf{S}_0$ by $m \mapsto \overline{m}$, where $\overline{m} = m \smallsetminus \ell . m$. Given $\overline{m}$ and $X \notin \overline{m}$, let $Z = m . \ell$. Then $\overline{XZ} \parallel \overline{m}$.

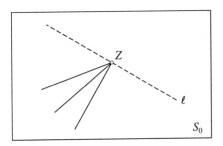

Figure E.5. A pencil of parallels.

(*b*) By Proposition 2.2, the completion of $\mathbf{S}_0$ is a projective plane **P**. Define an isomorphism of projective planes $\mathbf{P} \to \mathbf{S}$ by

$$x \mapsto \begin{cases} x & \text{if } x \in \mathbf{S}_0; \\ m.\ell & \text{if } x = P_{[m]}. \end{cases}$$

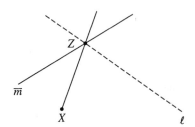

Figure E.6. $\overline{m} \parallel \overline{XZ}$.

2.3 (*b*) If there were a projective plane π of fewer than seven points, then, given a line ℓ, $\pi \smallsetminus \ell$ is an affine plane (by Exercise 2.2) with fewer than four points (by axiom P4).

195

2.4 All lines have the same number of points, since there is a one-to-one correspondence, called a perspectivity, between the ranges of points on two lines (refer to Figure 5.8). Also, a range of points (i.e., the set of points on a line) is in one-to-one correspondence with a pencil of lines (refer to Figure E.7). All the points of the projective plane are contained in a pencil of $n + 1$ lines at P. So we count points as follows: through the first line, we count $n + 1$ points, through the next n points, since P is not to be counted twice, and the same for the others. Hence there are $n^2 + n + 1$ points.

2.10 (*a*) By Exercise 2.4, each line ℓ in π has $n + 1$ points. It follows that each pencil $[P]$ has $n + 1$ lines by the simple bijection, $\ell \mapsto [P]$, $P \notin \ell$, given by

$$X \mapsto XP \qquad (X \in \ell)$$

where the inverse mapping is given by $m \mapsto m.\ell$.

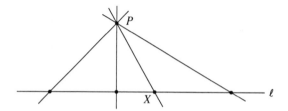

Figure E.7. An elementary correspondence.

2.11 First, we provisionally define (planar) projective geometry as those propositions about geometric figures that retain their meaning and validity after central projection between arbitrary planes (neither passing through the center of projection). Define affine geometry as those geometric propositions that retain their meaning and validity after parallel projection between arbitrary planes (neither containing a line of projection). Lastly define Euclidean geometry as those geometric propositions that retain their meaning and validity after parallel projections between parallel planes (neither containing a line of projection).

   Then each of the geometric figures are meaningful to Euclidean geometry, only (*b*), (*c*), (*d*), (*e*), (*g*), and (*h*) in affine geometry, and only (*a*), (*e*) in projective geometry.

3.11 Suppose π is a projective plane, in which the converse of P5 holds: i.e., each pair of axially perspective triangles is centrally perspective. Suppose triangles ABC and $A'B'C'$ are perspective from the point O. Show that the points $P = AB.A'B'$, $Q = AC.A'C'$, $R = BC.B'C'$ are collinear.

5.2 We compute in the xy-plane. Suppose A, B, C, D are the harmonic points $(a, 0), \ldots, (d, 0)$. Let ℓ be the line $x = a$. Let m be the line $y = x - a$, and n be the line $x = c$. Then apply the construction of the fourth harmonic point given in Proposition 5.7 and Figure 5.4. You will use the point-slope formula several times in order to get a relation among a, b, c, d.

5.7 (*a*) Apply the law of sines to the triangles OAC, OAD, OBC, and OBD, and the angles $\angle 1 - \angle 4$. For example,

$$\frac{\sin \angle 5}{OC} = \frac{\sin \angle 1}{AC}.$$

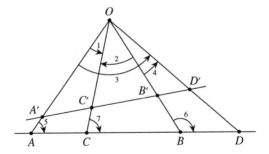

Figure E.8.

Eliminate the signed lengths OC and OD from your equations.

(b) The same trigonometric methodology as in (a) may be applied to the line $A'B'$.

(c) There is the special case of $O =$ ideal point to consider. By a theorem of Proclus, there exists a constant k such that the signed lengths $A'C' = k(AC), A'D' = k(AD), B'D' = k(BD), B'C' = k(BC)$.

5.8 (a) Inscribed angles in a circle that subtend the same arc have equal measure. (In fact, the measure is one half the radian length of the arc.)

(b) Each conic section is a central projection of a circle. Now apply (a) and Exercise 5.7.

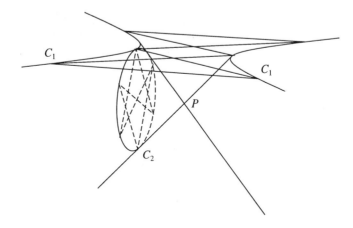

Figure E.9. The conic section C_1 centrally projects (via P) into the circle C_2.

5.15 Assign coordinates a, b, c, d, e. Apply Exercise 5.2 to get a relation among these variables. Show algebraically that $e = d$. The dual states the uniqueness of a fourth concurrent line in a fixed cross-ratio. By applying Exercise 5.7 to points of intersection on a transversal line, (a) will imply (b).

6.5 Suppose that $ABCX \barwedge A'B'C'X'$ and $ABCX \barwedge A'B'C'X''$. Assign coordinates a, b, c, x, x', x'' etc., on the x-axis. Recall that

$$R_x(A, B; C, D) = \frac{a-c}{a-d} \frac{b-d}{b-c}.$$

Apply Exercise 5.7 to get two equalities of cross ratios. Conclude from Exercise 5.15 that $x' = x''$.

6.10 (a) Let $S = C'Q.A'P$ and $T = C'R.A'Q$. Show that $PQ \parallel QR$ by showing that $\angle TRQ = \angle A'PQ$.

6.14 $\triangle A_1 X_1 P_1 \overset{\ell_2}{\barwedge} \triangle A_3 X_3 P_3$, where the notation means that the triangles are in perspective from the line ℓ_2.

6.20 $A = b.a', B = a.b.c, C = b.c'; A' = c.b', B' = a'.b'.c', C' = a.b'.$

8.12 (c) Since a field automorphism of $\mathbf{Z}_p$ satisfies $\alpha(\bar{1}) = \bar{1}$, it follows that $\alpha(\bar{2}) = \alpha(\bar{1} + \bar{1}) = \bar{2}$, so α is the identity. By Theorem 8.18, $\mathrm{Aut}\,\mathbf{P}^2(\mathbf{Z}_p) = \mathrm{PGL}(2, \mathbf{Z}_p)$. By Theorem 8.13, an automorphism ϕ of $\mathbf{P}^2(\mathbf{Z}_p)$ is completely determined by its values $\phi(X_1)$, $\phi(X_2)$, $\phi(X_3)$, and $\phi(X_4)$, on the vertices of the complete quadrangle $X_1 X_2 X_3 X_4$. $\phi(X_1)$ is any of $p^2 + p + 1$ points. $\phi(X_2)$ is any of the remaining $p^2 + p$ points. $\phi(X_3)$ is any point not on the line through $\phi(X_1)$ and $\phi(X_2)$: i.e., any of p^2 points. $\phi(X_4)$ is any point not on the triangle through $\phi(X_1)$, $\phi(X_2)$, and $\phi(X_3)$: i.e., any of $p^2 - 2p + 1$ points. Hence, ϕ is any of $(p^2 + p + 1)(p^2 + p)p^2(p - 1)^2$ possibilities. Check this against Section 4.2 when $p = 2$.

8.14 Let $X = \ell.AC$ and $Y = \ell.BC$, and define similarly $X' = \ell'.A'C'$, $Y' = \ell'.B'C'$, then A, B, X, Y are four points, no three collinear, and similarly for A', B', X', Y', so by Theorem 8.13 there is an automorphism ϕ of $\mathbf{P}^2(R)$ sending A, B, X, Y into A', B', X', Y'. Then clearly ϕ sends ℓ into ℓ' and C into C'.

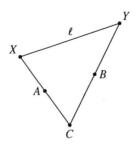

Figure E.10.

A.2 $x^2 + y^2 = 1$

References and Further Reading

[Artin]

E. Artin. *Geometric Algebra*. Wiley and Sons, New York, 1957.
One of the original sources for the theory in Chapter 9.

[Baer]

R. Baer. *Linear Algebra and Projective Geometry*. Academic Press, New York, 1952.
One of the original sources for the theory in Section 8.2.

[Bell]

Eric T. Bell. *Men of Mathematics*. Simon and Schuster, New York, 1965.
Biographies of Pascal, Poncelet, Lobatchevsky and others.

[Birkhoff]

G. Birkhoff. *Lattice Theory*. 3rd edition, Colloquium Publications, vol. 25, American Mathematical Society, Providence, 1967.
Interesting reading on lattices.

[Blumenthal]

L.M. Blumenthal. *A Modern View of Geometry*. Dover Publications, 1980.
Starts with a clear exposition of logic, axiomatics, consistency and independence with examples from geometry. Ternary rings and the basic postulational systems in Euclidean and non-Euclidean geometry.

[Chapman]

R.J. Chapman. *An Elementary Proof of the Simplicity of the Mathieu Groups M_{11} and M_{23}*. American Mathematical Monthly, vol. 102, no. 6 (1995), 544–545.
This article may be understood after a study of Chapter 2 in [Herstein]. Theorem 1 will tell you that the 168 element group $\mathrm{Aut}\,\mathbf{P}^2(\mathbf{Z}_2)$ of Section 4.2 is a simple group. What can you say about $\mathrm{Aut}\,\mathbf{P}^2(\mathbf{Z}_p)$?.

[Coxeter]

H.S.M. Coxeter. *Projective Geometry*. Springer Verlag, Heidelberg, 1987.
Synthetic projective geometry, with several chapters on polarity and conics.

[Coxeter]

H.S.M. Coxeter. *The Real Projective Plane, With an Appendix for Mathematica by George Beck*, 3rd. ed. Springer Verlag, 1993.
Characterizes the real projective plane with a complete set of axioms.

[Dörrie]

H. Dörrie. *100 Great Problems of Elementary Mathematics*. Dover Publications, New York, 1965.
The development of projective geometry and geometry in general can be traced here from a realistic problem-oriented viewpoint.

[Frankild-Kromann]

Anders Frankild and Matthias T. Kromann. Projektiv geometri og quaterniongeometri. *NATBAS Preprint series*, Roskilde University, 1993.
Account of projective geometry over division rings. In Danish.

[Hall]

Marshall Hall. *The Theory of Groups*. Chelsea, N.Y., 1972.
The last chapter of this book is entirely devoted to the subject of general projective planes and how they relate to groups and several classes of ring-like structures.

[Hartshorne]

Robin Hartshorne. *Foundations of Projective Geometry*, Lecture Notes, Harvard University. Benjamin Press, 1967.
The outline of the course we follow. The free projective plane is given in Chapter 2.

[Herstein]

I. N. Herstein. *Topics in Algebra*. Wiley and Sons, Inc., 1975.
An excellent undergraduate text on groups, rings, fields, modules; and linear algebra re-done from this viewpoint. After a reading of Chapter 7, you will be able to deduce that all finite Desarguesian planes are Pappian planes with $p^{2n} + p^n + 1$ points, for some prime power p^n.

[Hilbert]

David Hilbert. *Foundations of Geometry*. Open Court, La Salle, IL, 1971, trans. by L. Unger of *Grundlagen der Geometrie* (first publ. 1899).
The original monograph on the role of Desargues' theorem and Pappus' theorem (or Pascal's theorem as Hilbert would have it) in coordinatizing affine space.

[Holland]

S.S. Holland, Jr. Orthomodularity in infinite dimensions; a theorem of M. Solèr. *Bulletin of American Mathematical Society*, **32** (2), 1995, 205–234.
A characterization is given of real, complex and quaternionic projective geometry with polarity.

[Hsiang]

Wu-Yi Hsiang. On the laws of trigonometries of two-point homogeneous spaces. *Annals of Global Analysis and Geometry*, 7 (**1**), 1989, 29–45.
Lays the fundaments for more research in trigonometry.

[Jacobson]

N. Jacobson. *Basic Algebra 1*. Freeman and Co., San Fransisco, 1974.

The basic theory of groups, rings and fields with many unusual applications. Includes a chapter on lattices that covers the fundamental theorem of projective geometry and the Jordan-Hölder-Dedekind theorem.

[Kaplansky]

Irving Kaplansky. *Fields and Rings*. 2nd ed., University of Chicago Press, 1972.

Galois theory of fields and noncommutative ring theory.

[Kline]

Morris Kline. *Mathematical Thought from Ancient to Modern Times*. Oxford University Press, 1972.

Historical chapters on projective geometry, one on algebraic geometry and one on foundations of geometry.

[von Neumann]

John von Neumann. *Continuous geometry*. Princeton University Press, New Jersey, 1960.

Gives a proof of the existence part of the fundamental theorem for complemented modular lattices.

[Rees]

E.G. Rees. *Notes on Geometry*. Universitext, Springer Verlag, 1983.

Treats Euclidean, projective and hyperbolic geometry at a high level using linear algebra, group theory, metric spaces and complex analysis.

[Reid]

Miles Reid. *Undergraduate Algebraic Geometry*. London Mathematical Society Student Texts, Cambridge University Press, 1988.

Bezout's theorem for pairs of conics, as well as chapters on cubic curves, varieties and applications.

[Rubin-Silverberg]

K. Rubin and A. Silverberg. A report on Wiles' Cambridge lectures. *Bulletin of American Mathematical Society*, **31** (1), 1994, 15–38.

The recent attack on Fermat's Last Theorem.

[Ryan]

P. Ryan. *Euclidean and Non-Euclidean Geometry: an analytic approach*. Cambridge University Press, 1986.

The classical planar geometries and their automorphism groups using only vector analysis and a little matrix theory.

[Samuel]

P. Samuel. *Projective Geometry*. Undergraduate Texts in Mathematics, Readings in Mathematics, translated by S. Levy, Springer Verlag, 1988.

A general n-dimensional treatment of projective geometry including conics, quadrics, polarities and their classification.

[Schwerdtfeger]

>   Hans Schwerdtfeger. *Geometry of Complex Numbers*. Dover Publications; first edition: University of Toronto Press, 1962.
>
>   *The geometry of conics and cross ratio over the complex numbers.*

[Seidenberg]

>   A. Seidenberg. *Lectures in Projective Geometry*. University Series in Undergraduate Mathematics, van Nostrand Company, Princeton, 1962.
>
>   *Well-written book with chapters on conics, axioms for n-space, as well as projective geometry as an extension of a basic course in Euclidean geometry.*

[Silverman-Tate]

>   J. Silverman and J. Tate. *Rational Points on Elliptic Curves*. Undergraduate Texts in Mathematics, Springer Verlag, 1992.
>
>   *Clear account of elliptic curves and their group of rational points with an application of H. Lenstra to factorization of large integers. Appendix on projective geometry with the outline of a proof of Bezout's theorem.*

[Wylie]

>   C.R. Wylie, Jr. *Introduction to Projective Geometry*. McGraw-Hill Book Company, New York, 1970.
>
>   *Chapters on involutory hexads, the Cayley-Laguerre metrics and subgeometries of the real projective plane. Incidence table for a finite non-Desarguesian geometry.*

[Yale]

>   P. Yale. *Geometry and Symmetry*. Dover Publications, New York, 1988; first edition: Holden-Day, 1968.
>
>   *The similarities of n-dimensional projective geometry and the planar case is clearly seen in this book. Geometry from the transformation group viewpoint.*

Index of Notation

Set theory and logic
∀   for all, 3
∃   there exists, 3
⟹   implies, 3
⟺   if and only if (iff), 3
∅   empty set, 3
$X \smallsetminus B$ complement of B in X, 3
$\#(S)$ cardinality of S, 3

Algebra
$[x]$ equivalence class of x, 3
$a \equiv b \pmod{n}$ equivalence modulo n, 76
$\cong$ isomorphism, 32
$\mathrm{Ker}(\phi)$ kernel of homomorphism, 42
G/N factor group, 120
$\langle g \rangle, \langle S \rangle$ generated group, 38, 39
$Z(G), Z(R)$ center of group or ring, 80
gH, Hg left and right cosets, 37
H_x stabilizer subgroup, 37
β_x orbit of x, 38
$(i_1 i_2 \ldots i_r)$ r-cycle, 40
$\mathrm{char}\, F$ characteristic of division ring, 77
$\mathrm{Aut}\, G, \mathrm{Aut}\, R$ automorphisms of group or ring, 81
$\mathrm{Inaut}\, G, \mathrm{Inaut}\, R$ inner automorphisms, 99
$H \rtimes K$ semi-direct product, 110

Synthetic geometry
$PQ, P \cup Q$ line through P and Q, 3
$\ell . m, \ell \cap m$ intersection of lines, 3
$\ell \parallel m$ parallel lines, 3
ℓ_∞ line at infinity, 15
$P_{[\ell]}$ point at infinity, 14
m_n configuration of m points, n points per line, 25

π^* dual projective plane, 29
$ABC \overset{O}{\barwedge} A'B'C'$ perspectivity, 52
$ABC \barwedge A'B'C'$ projectivity, 53
$\mathrm{H}(A,B;C,D)$ harmonic quadruple, 48
$\triangle ABC \sim \triangle DEF$ similar triangles, 145

Analytic geometry
$\mathbf{A}^2(R)$ affine plane over R, 7
$\mathbf{P}^2(R)$ projective plane over R, 18, 87
$\mathbf{P}^3(R)$ projective space over R, 88
$\mathbf{P}^n(R)$ projective n-space over R, 89
S^2 standard 2-sphere, 19
$\mathrm{dist}(P,Q)$ Euclidean distance, 11
$R_x(A,B;C,D)$ cross ratio, 57
$|x|$ Euclidean norm, 78

Specific groups
S_n permutation of n letters, 39
$\mathrm{Perm}\, S$ permutations of S, 36
$\mathrm{Aut}\, \mathbf{A}, \mathrm{Aut}\, \mathbf{C}, \mathrm{Aut}\, \mathbf{P}$ automorphisms of plane or configuration, 6, 36
$\mathrm{Dil}\, \mathbf{A}$ dilatations of affine plane, 8
$\mathrm{Dil}_O(\mathbf{A})$ dilatations fixing O, 110
$\mathrm{Tran}\, \mathbf{A}$ translations of affine plane, 9
$\mathrm{Tran}_m(\mathbf{A})$ translations in direction m, 114
$\mathcal{H}(\ell, O)$ homologies with axis ℓ and center O, 138
$\mathcal{E}(\ell)$ elations with axis ℓ, 137
$\mathrm{PJ}(\ell)$ projectivities of a line into itself, 54
$\mathcal{R}(\ell)$ cross ratio preserving transformations, 129
$\mathrm{PGL}(1,R), \mathrm{PGL}(2,R)$ projective general linear groups, 93, 125
$\mathrm{PC}(\pi)$ projective collineations, 136

203

Index